Rugerri Toni Liong

Application of the cohesive zone model to the analysis of rotors with a transverse crack

Karlsruher Institut für Technologie

Schriftenreihe des Instituts für Technische Mechanik

Band 16

Application of the cohesive zone model to the analysis of rotors with a transverse crack

by
Rugerri Toni Liong

Dissertation, Karlsruher Institut für Technologie
Fakultät für Maschinenbau
Tag der mündlichen Prüfung: 30. September 2011

Impressum

Karlsruher Institut für Technologie (KIT)
KIT Scientific Publishing
Straße am Forum 2
D-76131 Karlsruhe
www.ksp.kit.edu

KIT – Universität des Landes Baden-Württemberg und nationales
Forschungszentrum in der Helmholtz-Gemeinschaft

KIT Scientific Publishing 2012
Print on Demand

ISSN: 1614-3914
ISBN: 978-3-86644-791-2

Application of the cohesive zone model to the analysis of rotors with a transverse crack

---

Zur Erlangung des akademischen Grades

Doktor der Ingenieurwissenschaften

der
Fakultät für Maschinenbau
Karlsruher Institut für Technologie (KIT),

genehmigte
Dissertation

von

M.Eng. Rugerri Toni Liong
aus Jakarta, Indonesien

Tag der mündlichen Prüfung: 30. September 2011
Hauptreferent: Prof. Dr.-Ing. Carsten Proppe
Korreferent: Prof. Dr.-Ing. habil. Jens Strackeljan

Vorwort

Die vorliegende Arbeit enstand während meiner Tätigkeit als Stipendiat am Institut für Technische Mechanik des Karlsruher Instituts für Technologie in den Jahren 2007 bis 2011.

Mein besonderer Dank gilt Herrn Prof. Dr.-Ing. Carsten Proppe für die großzügige Unterstützung und wissenschaftliche Förderung meiner Arbeit und für die Übernahme des Hauptreferates. Ganz besonders möchte ich mich auch für das entgegengebrachte Vertrauen und die wissenschaftliche Freiheit bedanken.

Herrn Prof. Dr.-Ing. habil. Jens Strackeljan vom Institut für Mechanik der Otto-von-Guericke Universität Magdeburg danke ich für die Übernahme des Korreferates und die wertvollen Anmerkungen. Dem Vorsitzenden des Prüfungsausschusses, Herrn Prof. Dr.-Ing. Kai Furmans vom Institut für Fördertechnik und Logistiksysteme des Karlsruher Instituts für Technologie gilt ebenfalls mein Dank.

Des Weiteren danke ich Herrn Prof. Dr.-Ing. Dr. h.c. Jörg Wauer vom Institut für Technische Mechanik des Karlsruher Instituts für Technologie und Herrn Prof. Dr. Ahmad A. Al-Qaisia von der University of Jordan, Amman, Jordanien für ihr Interesse an meiner Arbeit. Auch ihnen verdanke ich eine Reihe wertvoller Hinweise zur Verbesserung der Dissertation.

Außerdem möchte ich mich bei Herrn Prof. Dr.-Ing. Wolfgang Seemann, Herrn Prof. Dr.-Ing. habil. Alexander Fidlin, Herrn Prof. Dr.-Ing. Dr. h.c. Jens Wittenburg und Herrn Prof. Dr.-Ing. Walter Wedig bedanken, die in zahlreichen Diskussionen wichtige Impulse gaben.

Herrn Prof. Dr. Evgeny Glushkov von der Kuban State University, Russland, und Herrn Prof. Dr. Marko Todorov von der University of Rousse, Bulgarien, möchte ich für die schöne Zeit im gemeinsamen Arbeitszimmer und für Anregungen und Diskussionen danken.

Den Mitarbeitern, den Stipendiaten sowie Kollegen am Institut für Technische Mechanik möchte ich für die wunderbare Arbeitsatmosphäre danken. Ganz besonders danke ich Herrn Dipl.-Ing. Christoph Baum für die Korrektur des Promotionsvortrags und Herrn Dr.-Ing. Günther Stelzner für die Unterstützung bei der Softwareanwendung sowie Frau Gudrun Volz und Frau Dipl.-Ing. Heike Vogt für die Unterstützung beim Erstellen deutscher Unterlagen. Zudem möchte ich mich bei allen ehemaligen Mitarbeiter des Instituts für Technische Mechanik bedanken - es war eine tolle Zeit.

Herrn Dr. Heinrich Geiger und Frau Karin Bialas vom Katholischen Akademischen Ausländer-Dienst (KAAD) Bonn danke ich für die finanzielle Unterstützung meines Studiums in Deutschland. Ebenfalls möchte ich mich bei der katholischen Universität Atma Jaya, Indonesien, meiner Heimatuniversität, bedanken, die meine Familie und mich ebenfalls finanziell unterstützten.

Von tiefstem Herzen danke ich meiner lieben Ehefrau Juliana und meinen Töchtern Evelyn und Claresta. Mit viel Geduld und fortwährender Unterstützung begleiteten sie mich bei dieser Arbeit. Ohne Euch hätte mir die Kraft und die Ausdauer gefehlt.

Karlsruhe, den 28. November 2011
Rugerri Toni Liong

Für

meine liebe Ehefrau Theresia Juliana

und

meine lieben Töchter Evelyn and Claresta

Abstract

Ein Riss im Rotor ruft eine lokale Steifigkeitsänderung hervor. Die vorliegende Dissertation ermittelt die Steifigkeitsänderung einer angerissenen Welle. Dazu wird ein Kohäsivzonenmodell eingesetzt. Das Modell wurde für die erste Rissöffnungsmode bei ebenem Verzerrungszustand in Abhängigkeit der Mehrachsigkeit des Spannungszustandes (Triaxialität) entwickelt. Dafür wird ein elastisch-plastisches Kohäsivgesetzt verwendet. Aufgrund von Finite-Elemente (FE) Simulationen und experimentellen Ergebnissen des Rissfortschritts einer angerissenen Welle wird die Rissform als parabolische Form modelliert. Solange das Risstiefenverhältnis klein ist, lässt sich eine geradlinige Rissform ansetzen. Um lediglich den Einfluss des Risses abzubilden, wird ein starr gelagerter Rotor angenommen. Außerdem wird das Kohäsivzonenmodell bei einem eindimensionalen Kontinuumsrotor als FE Modell ausgeführt. Dafür werden zwei verschiedene FE Modelle vorgeschlagen und diskutiert, ein FE Modell mit lokaler Steifigkeit infolge des Risses, sowie ein FE Modell mit Elementen mit Breite Null aber mit kohäsiven Eigenschaften. Um die Gültigkeit des Rissformmodells abzuschätzen, werden die Ergebnisse mit dem Verlauf des Rissöffnungsmechanismus für einen Rotor aus FE- und Mehrkörpersimulationen verglichen. Es zeigt sich, dass das gradlinige Rissformmodell für kleine Risstiefen akzeptable Ergebnisse liefert. Für eine realistischere und genauere Modellierung des Rissöffnungsmechanismus ist das Kohäsivzonenmodell geeignet.

List of Symbols and Abbreviations

List of Symbols

a	length of edge crack in a plate
a	crack depth in a shaft
a_{eff}	effective length of edge crack
a_0	initial crack length in a plate
a_1, a_2, a_3, a_4	complex coefficients of fourth order differential equation
a/d	relative crack depth (crack depth ratio)
a/R	ratio of crack depth to the shaft radius
a_ξ, a_η	coefficient for perturbation method
A	cross section area
$\mathbf{A}$	system matrix
A_{cr}	cracked area of the cross section
A_{ucr}	rest uncracked area of the cross section
b	width of thin plate
b_ξ, b_η	coefficient for perturbation method
c	length of the plastic zone
c	damping coefficient
$\mathbf{c}$	damping matrix
c_L	local stiffness
$\bar{c}$	dimensionless local flexibility of the cracked section
C	non-dimensional material constant
c_1, c_2, c_3, c_4	arbitrary constants
$\dot{d}$	scalar damage variable
d	diameter of the shaft
d_{ij}	flexibility element of matrix

$\tilde{d}$	distance from an axis to the centroid of the cross section
D_{ij}	constitutive operator of the crack interface
D_{in-d}	inner diameter of the disk
D_{out-d}	outer diameter of the disk
D_θ	parameter of diameter of the shaft
e	eccentricity of the unbalance mass
E	modulus of elasticity (Young's modulus)
E_{coh}	cohesive energy
E_{el}	elastic strain energy
E_{inel}	inelastic energy
E_{kin}	kinetic energy of the body
E_{pl}	plastic dissipative energy
$f(t)$	function representing the opening/closing effect (breathing steering function)
$f_{coh}(t)$	breathing steering function modeled by cohesive zone model
f_{kx}, f_{ky}	elastic force in x- and y-direction
F	applied force (load)
$\mathbf{F}^g$	gravity (weight) force vector
$\mathbf{F}_k$	vector of elastic force due to deformation of the shaft
$\mathbf{F}^{ub}$	unbalance (centrifugal) force vector
$F\left(\frac{\alpha}{\alpha'}\right)$	factor of crack geometry
g	gravity acceleration
$\mathbf{g}$	gyroscopic matrix
g_ξ, g_η	flexibility of the cracked shaft in ξ- and η-axes
$g_{\xi\eta}$	cross-coupled flexibility of the cracked shaft
$\hat{g}_\xi, \hat{g}_\eta$	flexibility due to crack in ξ- and η-axes
$\hat{g}_{\xi\eta}$	cross-coupled flexibility due to crack
G	shear modulus
G_I	strain energy release rate (cohesive energy)
G_{IC}	critical strain energy release rate (maximum cohesive energy)
h, h_1	damping coefficient for perturbation method
I	area moment of inertia of a circular section about its centroid
$\tilde{I}_{cr}$	area moment of inertia of a cracked shaft

I_d	diametral moment of inertia about axis perpendicular to rotor axis
I_p	mass polar moment of inertia about the rotor axis
$I_X,\ I_Y$	area moment of inertia of the cracked element about its centroidal axes
$\bar{I}_X,\ \bar{I}_Y$	area moment of inertia about the X- and Y-axes
j	imaginary unit
J	strain energy density function (J-energy)
J_{IC}	critical J-energy (fracture energy)
k	shaft stiffness
$\mathbf{k},\ \mathbf{K}$	stiffness matrix
$\mathbf{k}_{cr}$	crack element stiffness matrix
$\mathbf{k}_{coh}$	local element stiffness matrix at cohesive crack
$\mathbf{K}_{coh}$	cohesive element stiffness matrix
k_ξ, k_η	stiffness of the cracked shaft in ξ- and η-axes
$k_{\xi\eta}$	cross-coupled stiffness of the cracked shaft
$K,\ K_m,\ K_n$	material constant
K_p	penalty stiffness (interface stiffness)
K_I	stress intensity factor for mode-I
K_{IC}	critical stress intensity factor for mode-I
$K_{I\,eff}$	effective stress intensity factor
l_{lig}	length of uncracked ligament
l_e	length of element
L	length of the shaft
L_d	length of the disk
m	parameter for perturbation method
$m,\ m_d$	mass of the disk
m_s	mass of the shaft
m_{ub}	mass unbalance
$\mathbf{m}_R$	rotational inertia matrix
$\mathbf{m}_T$	translational inertia matrix
n	natural number, $n = 1, 2, 3, \cdots$
n	material exponent constant (strain hardening exponent)
n	parameter for perturbation method

$\mathbf{N}$	shape function matrix
p, q	geometric constants
p_1, p_2, p_3, p_4	arbitrary constants
q	normalised coordinate
$\mathbf{q}$	functions of the displacements at the nodes matrix in stationary coordinate
$\mathbf{q}^{rot}$	functions of the displacements at the nodes matrix in rotating coordinate
Q	applied force (load)
$r_c, \dot{r}_c, \ddot{r}_c$	position, velocity and acceleration in rotating coordinate
r_0	radius of gyration
r_1, r_2, r_3, r_4	arbitrary constants
r_σ	stress ratio
$r_{\sigma\,cr}$	critical value of biaxiality stress ratio
R, r	radius of the shaft
R_{in-d}	inner radius of the disk
R_{out-d}	outer radius of the disk
S	non-dimensional multiplicative factor of the elastic strain at yield
t	plate thickness
$\mathbf{t}$	exterior surface traction vector
T	torque
T	kinetic energy
T_s	kinetic energy of the shaft
T_d	kinetic energy of the disk
$\mathbf{T}$	coordinate transformation matrix
u	displacement in x-direction
$\hat{u}$	additional deflection
u_i	displacement points of the interface
U	potential energy
U	total strain energy
U_c	strain energy due to crack
U_s	potential energy of the shaft
U_0	strain energy of the uncracked shaft
v	displacement in y-direction

$\mathbf{v}$	velocity field vector
w	crack width
W	external work
x	stationary coordinate; horizontal direction
x	distance ahead of the crack in a plate
$\overline{x}_i$	coordinate of the deformed interface
X_i	coordinate of the undeformed interface
y	stationary coordinate; vertical direction
y_i	deflection at the rotor in vertical direction
$y(z)$	mode shape function at the rotor in vertical direction
z	stationary coordinate; axial coordinate
α	crack depth as function of the angle of rotation
α_c	proportional damping factor to mass
α_m	material parameter
α_p	parameter for penalty stiffness
$\beta,\ \beta_1$	ratio between location and length of the shaft
β_c	proportional damping factor to stiffness
γ	shear deformation
δ	angle of unbalance mass
δ	separation
$\dot{\delta}$	cohesive separation rate
$\delta_c,\ \delta_{sep}$	critical separation
$\delta_i,\ \delta_j$	displacement jump of element
δ_m	deformed interface
δ_0	separation until crack initiation
δ_1	linear separation limit in traction-separation law
δ_2	strain hardening separation limit in traction-separation law
$\delta_\xi,\ \delta_\eta$	parameter for perturbation method
$\Delta^\xi,\ \Delta^\eta$	total deflection in ξ- and η-axes
ε	strain
ε	eccentricity

$\dot{\varepsilon}_{el}$	elastic strain rate
$\dot{\varepsilon}_{pl}$	plastic strain rate
$\bar{\varepsilon}^{pl}$	effective plastic strain at fracture initiation
$\varepsilon_\xi,\ \varepsilon_\eta$	parameter for perturbation method
ζ	damping ratio
η	rotating coordinate
$\dot{\eta},\ \ddot{\eta}$	rotating speed and acceleration in rotating coordinate
$\theta,\ \dot{\theta},\ \ddot{\theta}$	angle of rotation, rotating speed and rotating acceleration
Θ_m	rotation tensor
κ	shear coefficient (form factor for flexural vibrations of beam)
λ	eigenvalue or root of the characteristic equation
μ	ratio of crack depth to the shaft radius
ν	Poisson's ratio
ξ	rotating coordinate
$\dot{\xi},\ \ddot{\xi}$	rotating speed and acceleration in rotating coordinate
ρ	material density
$\rho,\ \dot{\rho},\ \ddot{\rho}$	position, speed and acceleration in rotating coordinate
σ	normal stress
σ	traction
σ_j	element of traction
$\sigma_1,\ \sigma_2,\ \sigma_3$	principal stresses
σ_a	applied stress
σ_{eq}	effective stress (von Mises equivalent stress)
σ_{max}	maximum stress; cohesive strength
σ_m	mean normal stress; hydrostatic stress
σ_U	ultimate strength of material
σ_Y	yield strength of material
ς	nondimensional in z-coordinate
τ	parameter of time for perturbation method
τ_{max}	shear strength
ϕ	angle of rotation of element
$\varphi_\xi,\ \varphi_\eta$	parameter for perturbation method

χ	triaxiality of the stress state
χ_{eff}	effective triaxiality parameter
χ_{el}	elastic triaxiality of stress state
χ_{pl}	extreme or plastic triaxiality of stress state
χ_{sat}	saturation limit of the triaxiality parameter
ψ	free energy density per unit surface of the interface
ψ_0	free energy per unit surface of the interface
ω	angular frequency
ω_n	natural frequency
$\omega_\xi,\ \omega_\eta$	natural frequency in ξ- and η-axes
Ω	rotational speed of the rotor
Ω_{cr}	critical speed of the rotor
ϑ	ratio of the beam bending stiffness to the shear stiffness

List of Abbreviations

CTOD	Crack-tip opening displacement
CZM	Cohesive zone model
DCB	Double cantilever beam
EPFM	Elastic-plastic fracture mechanics
FE	Finite element
FEM	Finite element method
GTN	Gurson-Tvergaard-Needleman
LEFM	Linear-elastic fracture mechanics
MBS	Multi-body simulation
MMB	Mixed-mode bending
MPC	Multi-point constrain
SIF	Stress intensity factor
TSL	Traction-separation law

Contents

List of Symbols and Abbreviations   **vii**

1 Introduction   **1**
   1.1 Motivation for the research 2
   1.2 Literature review 6
      1.2.1 Cracked rotor 6
      1.2.2 Cohesive zone model 10
   1.3 Objectives 12
   1.4 Structure of the dissertation 14

2 Cohesive zone models   **17**
   2.1 Fracture process zone 17
   2.2 Ductile and brittle fracture 20
   2.3 Cohesive zone model 21
      2.3.1 Basis of the models 21
      2.3.2 Location of crack tip 22
      2.3.3 Kinematics and constitutive relation of cohesive zone models 23
   2.4 Finite element implementation 24
      2.4.1 One cohesive element test 25
      2.4.2 Stress distribution on a cracked plate 28
         2.4.2.1 Linear elastic fracture mechanics (LEFM) 28
         2.4.2.2 Elastic-plastic fracture mechanics (EPFM) 29
      2.4.3 Simulation of delamination 33
   2.5 Energy balance concept 34
      2.5.1 Simulation of crack growth 36
      2.5.2 Energy distribution in purely elastic material 38
      2.5.3 Energy distribution in elasto-plastic material 39
   2.6 Triaxiality dependent cohesive zone model 42
      2.6.1 Introduction 42
      2.6.2 Constitutive behaviour of the continuum (undamaged material) . . 43
      2.6.3 Formulation of triaxiality dependent Model 45
      2.6.4 Traction-Separation Law 47
   2.7 Implementation of the cohesive model for a cracked shaft 53
      2.7.1 Classification of cracks 53
      2.7.2 Breathing crack under rotating load 53
      2.7.3 Breathing versus open crack: change in second moment of area . . . 54

2.7.4 Breathing versus open crack: change in local stiffness 57

3 Rotor with breathing transverse crack   **61**
3.1 Breathing crack modeling 61
    3.1.1 Breathing steering functions 64
    3.1.2 Breathing crack shapes 65
3.2 Dynamics of cracked rotors 70
    3.2.1 Mathematical formulation 70
    3.2.2 Equations of motion of the de Laval rotor with breathing crack . . . 74
3.3 Stiffness estimation based on linear elastic fracture mechanics 75
3.4 Stiffness estimation based on cohesive zone model 79
3.5 Breathing crack with large crack depth ($a/d > 0.2$) 89

4 Stability analysis of a rotor with a transverse breathing crack   **93**
4.1 Stability of rotor systems 93
4.2 Instability due to parametric excitation 94

5 Finite element model of the cracked shaft   **103**
5.1 Model of the rotor supported by rigid bearings 104
    5.1.1 Shaft model 104
    5.1.2 Disk model 109
5.2 Model of the cracked shaft 111
    5.2.1 Published models of cracked shaft 111
    5.2.2 Model based on asymmetric area moments of inertia 113
    5.2.3 Finite element results 115
       5.2.3.1 Natural frequencies 115
       5.2.3.2 Mode shapes 116
       5.2.3.3 Effect of crack depth 116
       5.2.3.4 Comparison of results 118
    5.2.4 Model based on zero-thickness element 121
5.3 Cracked rotor supported by rigid bearings with disk 125
    5.3.1 Finite element results 125
    5.3.2 Comparison of results 125

6 Breathing crack simulation   **129**
6.1 Finite element model of flexible cracked shaft 131
6.2 Dynamic behaviour of rotating flexible cracked shaft 133
    6.2.1 Flexible cracked shaft loaded by weight only 133
    6.2.2 Flexible cracked shaft loaded by weight and unbalance 137
       6.2.2.1 Case 1: Deep crack $a/d = 0.5$, large unbalance mass . . . 137
       6.2.2.2 Case 2: Deep crack $a/d = 0.5$, small unbalance mass . . . 138
       6.2.2.3 Case 3: Shallow crack $a/d = 0.1$, large unbalance mass . . 139
6.3 Breathing mechanism in finite element simulation 141
    6.3.1 Breathing crack under rotating loading 141
    6.3.2 Breathing crack during rotation of the shaft 145

6.4   Validation of the breathing crack model 150

7   Conclusions   **153**
7.1   Major results 153
7.2   Recommendations and further analysis 156

Appendix   157

A   Analytical methods for rotating shafts with open crack   **157**
A.1   Timoshenko beam theory for rotating shaft 157
A.2   Dunkerley's equation 162
A.3   Rayleigh's method 164

List of figures   166

List of tables   171

References   173

1 Introduction

Shafts are amongst components subjected to perhaps the most arduous working conditions in high performance rotating equipment used in process and utility plants. Although usually quite robust and well designed, shafts in operation are sometimes susceptible to serious defects that develop without much apparent warning. They are prime candidates for fatigue cracks because of the rapidly fluctuating nature of bending stresses, the presence of numerous stress raisers and possible design or manufacturing flaws. The mechanism leading to cracks in rotors, which are considered to be of primary importance, include high-cycle fatigue, low-cycle fatigue, severe heat-related stress cracks, corrosion-related stress cracks, creep-related high temperature alloy rotor cracks and probably other mechanism, not yet well identified shown in Figure 1.1.

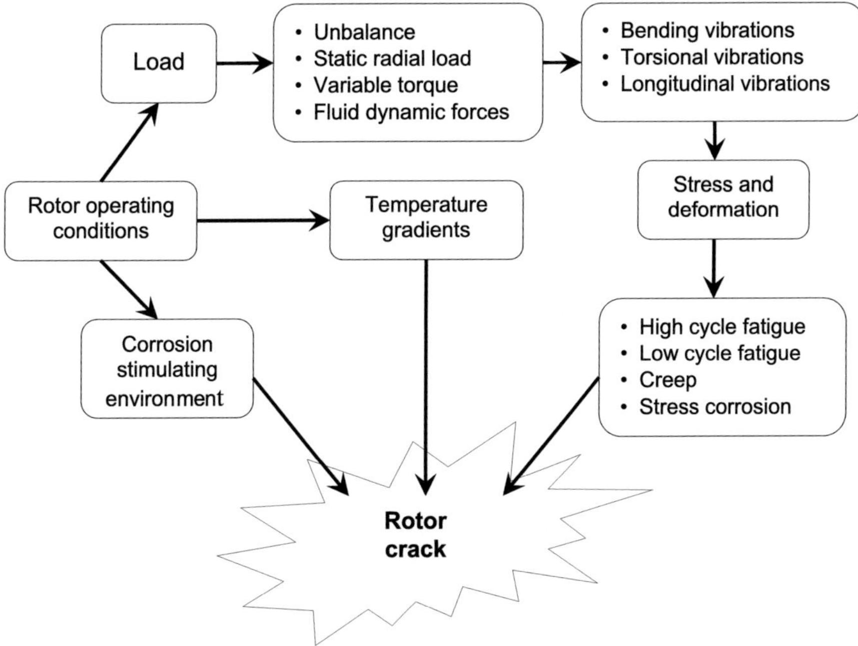

Figure 1.1: Physical phenomena leading to rotor cracks [91]

1.1 Motivation for the research

Presence of cracks in shafts is considered as one of the most important factors that limit the safe and reliable operation of rotating machinery. Multiple catastrophic failures of machines, caused by cracked rotors, have increased the interest in early detection of rotor cracks. In general, non destructive testing is used in inspection intervals to prevent such failures, but recently vibration analysis has received much attention in trying to continuously monitor the condition of machines. In particular, the crack depth and its location along the rotor are important parameters, which indirectly would provide information that can be used in early detection of cracks.

Until today, the greatest difficulty in crack detection and identification remains the quantitative evaluation of the crack parameters and distinction between a developing crack from other faults such as imbalance, shaft misalignment, asymmetric shaft, bearing failure, looseness of bolts and nuts or a range of other nonlinearities [70], [136]. The key issues in developing an accurate modeling technique of a cracked rotor are the reduced stiffness of the cracked cross section, the variation of stiffness over one revolution due to opening and closing of the crack and the complexity in geometry of the rotor, in particular in the region of the developing crack.

Cracks perpendicular to the shaft axis are known as transverse cracks. These are the most common and most serious as they reduce the cross-section and thereby weaken the rotor. Most past and current research focuses on the detection of such cracks. Cracks perpendicular to the shaft introduce a local flexibility in the stiffness of the shaft due to strain energy concentration in the vicinity of the crack tip [116]. Due to shaft self-weight and the rotation of the rotor, the crack opens and closes during a complete revolution of the rotor. Hence, the stiffness of the shaft varies. Cracks which open when the affected part of the material is subjected to tensile stresses and close when the stress is reversed are known as breathing cracks. The stiffness of the component is most influenced when under tension. Usually, shaft cracks breathe when crack sizes are small, running speeds are low and radial forces are large [47]. Most theoretical research efforts are concentrated on transverse breathing cracks due to their direct practical relevance.

The breathing mechanism of the crack comes from the fact that the static deflection of the shaft is much greater than the deflection due to the dynamic response of the cracked rotor. The presence of a transverse crack in a structural member introduces local flexibility, which can be mathematically modelled by a local compliance (flexibility) matrix. Theoretically, coefficients of the compliance matrix of the crack in bending can be computed based on available expressions for stress intensity factors (SIFs) and associated expressions of the strain energy density function over the opened crack surface by using linear elastic fracture mechanics (LEFM). These changes, in turn, affect the dynamics of the system: natural frequency of the vibrations and the amplitudes of forced vibrations are changed. There were a number of publications in the literature dealing with the influence of a transverse crack on the vibration of a rotating shaft.

Generally, three different crack states have been introduced in order to investigate the dynamic behaviour of cracks in rotors. The first one is the opening crack: if the vibration amplitudes due to any out-of-balance forces acting on a rotor are greater than the static deflection of the rotor due to gravity, then the crack will remain open (or closed) depending on the size and location of the unbalance masses. When the crack is assumed to remain open during the revolution, the system does not differ from the general case of the anisotropic rotor. The rotor is then asymmetric and this condition can lead to stability problems [107]. The other one is the switching crack (also known as hinge model) where the crack is assumed to open and close fully following a square pulse function. This behaviour is connected to a rather small crack, while a deep crack could be associated with a harmonic variation. Another one is the regularly periodic breathing crack for heavier rotors, where the vibration due to any out-of-unbalance forces acting on a rotor is less than the deflection of the rotor due to gravity. This crack may open and close during rotation.

Breathing crack and switching crack models differ clearly from each other. For the breathing crack model, the open part of the crack continuously changes with the shaft rotation, thereby accounting for the partial open and close state of the crack. In contrast, in the switching crack model, the crack is assumed to be in either fully open state or fully closed state, partial opening and closing of the crack is not accounted for. In real rotors, the stiffness variation is likely to be gradual and the assumption of abrupt sudden stiffness switching is not appropriate [104]. In general, two different approaches can be distinguished to model the behaviour of a breathing crack in rotating structures. The first approach is based on the fact that the presence of a crack in a rotating shaft reduces the stiffness of the structure, hence reduces the natural frequencies of the original uncracked shaft. The second approach is the response-dependent breathing crack model, which estimates the status of the crack closure using a force acting on the crack section, and thereby evaluates the stiffness of the cracked section.

During the last four decades, great attention has been paid by several scientists to the diagnosis of cracks in rotating machinery. The challenge of modelling a crack is one of the most significant issues in this area. The theory of strain energy release rate and SIFs combined with rotordynamics built the foundation of the dynamic analysis for crack rotors based on LEFM. However, this theory has two major limitations, namely, it can only be used if there is an initial crack and the fracture process zone must be small compared to the dimensions of the shaft. Many researchers have used the theory of strain energy release rate combined with rotordynamics in order to investigate the dynamic behaviour of breathing crack on the rotor. Due to geometrical complexity, some simplifications had been made for the crack profile, such as straight-edged, circular and elliptical crack model to analyze such problems. Many works using cross-section of a straight-edge crack can be found in the literature. Since there are not available stress intensity factors for an edge crack in a rotating circular cross section, the shaft is considered to be a sum of elementary independent rectangular strips and the stress intensity factors, known for the rectangular beam are integrated along the crack tip.

In the above crack models, the crack tip is supposed to be formed by the boundary between the cracked areas and the uncracked areas for the regions in which the breathing crack is open, which is correct. However, the SIF will not appear at the boundary between the closed cracked areas and the open cracked areas. Furthermore, many approaches assume stress and strain distributions with same values along directions parallel to the applied bending moment axis (as they are in rectangular cross sections), and no interaction between parallel rectangular strips in which the circular cross section have been divided. The cracked cross section is no longer planar, but distorted. This is not taken into account by the LEFM approach [9].

LEFM has proven a useful tool for investigating the dynamics of a cracked rotor when the nonlinear zone ahead of the crack tip is negligible. However, for ductile materials, the size of the nonlinear zone due to plasticity or microcracking is not negligible in comparison with other dimensions of the cracked geometry. In addition, the presence of an initial crack is needed for LEFM to be applicable. This means that shafts with notches but without crack can not be analyzed using LEFM. Furthermore, in reality, neither the idealized sharp crack nor the linear elastic material does exist. Although there are many works on modelling of cracked rotor based on LEFM reported in the literature, they are still until now not completely closed. The direct motivation of this research stems from the alternate method in order to model the dynamics of a cracked rotor instead of LEFM.

Nowadays, the cohesive zone model (CZM), a model which can deal with the nonlinear zone ahead of the crack tip due to plasticity or microcracking, has been widely used instead of fracture mechanics based on SIFs. The CZM describes material failure on a more phenomenological basis (i.e. without considering the material microstructure). The general advantage of CZM when compared to LEFM is that the parameters of the respective models depend only on the material and not on the geometry. This concept guarantees transferability from specimen to structure over a wide range of geometries.

The origin of the cohesive zone concept can be traced back to the strip yield model in which the narrow zone of localized deformation ahead of the crack tip was substituted by cohesive traction between the bounding surfaces. The constitutive behaviour which causes the cohesive elements to open and eventually to fail is described by the so called traction-separation law (TSL). It relates the traction vector to the displacement jump across the interface. Various functions have been chosen for the shape of the TSL. Common to all shapes is that they contain two model parameters, namely the maximum traction sustainable by the element and a critical separation at which the element finally fails. The energy dissipated by the element until total failure is derived as the integral of the TSL. However, the TSL depends on the stress state, which can be characterised by the triaxiality, which is the hydrostatic stress divided by the von Mises equivalent stress. The effect of triaxiality on conventional cohesive parameters is well predicted as peak stresses are known to increase while the cohesive energy decreases with triaxiality.

The CZM, originally applied to concrete and cementitious composites, can be used with success for other materials, such as ductile materials. This model is able to adequately predict not only the behaviour of cracked structures but also of uncracked structures including notches. Furthermore, it avoids non-realistic stress singularities near the crack tip. More powerful computer programs and better knowledge of material properties may widen its potential field of application. The other advantage of the CZM lies in the fact that it considers plasticity and that crack propagation can be easily included in the analysis. To the best of the author's knowledge, CZMs have been used to simulate the fracture process only in static structures and application of this model on a cracked rotor has not been studied yet.

There are various tools and methodologies to be applied to model a cracked rotor. Each methodology has specific advantages. Important questions with regard to modelling may be summarized as follows:

- Crack breathing mechanism plays an important role in analysis of dynamic behaviour of a cracked rotor. This phenomenon must be modelled accurately to analyze the crack in a rotor. Although there are many models published in literature, how accurate is the prediction model?

- The presence of a transverse crack in a structural member introduces local flexibility, which can be mathematically modelled by a local compliance (flexibility) matrix. Theoretically, coefficients of the compliance matrix of the crack in bending can be computed based on available expressions for SIFs and associated expressions of the strain energy density function over the opened crack surface by using linear elastic fracture mechanics. Nevertheless, direct application of the LEFM method is not possible, because solutions for the SIF, for a cylindrical shaft with an edge crack, are not available. Is there another technique which can be used to analyse such problems?

- The CZM was used by various researchers to predict the generation and propagation of a crack in structure. This propagation mechanism may be modelled so that the early failure of the structure may be detected. Can this model be applied in a dynamic problem such as a rotating shaft instead of LEFM?

- Finite element method (FEM) is a better choice and applied by various researchers to analyze the dynamic behaviour of a shaft having different kind of cracks, for example, transverse crack, two cracks, slant crack, and so forth. The crack element must be accurately discretized to depict the real behaviour of a cracked rotor. Can CZM along with FEM provide a contribution to the cracked rotor model?

1.2 Literature review

1.2.1 Cracked rotor

A breathing crack in the transverse direction of a shaft can cause dangerous damage in rotor dynamic systems. The breathing mechanism of the crack comes from the fact that the static deflection of the shaft is much greater than the deflection due to the dynamic response of the cracked rotor. A crack in the rotor causes local changes in stiffness. These changes, in turn, affect the dynamics of the system: natural frequency of vibrations and the amplitudes of forced vibrations are changed. The influence of a transverse crack on the vibration of a rotating shaft has been in the focus of many researchers. A comprehensive literature survey of various crack modelling techniques, system behaviour of cracked rotor and detection procedures to diagnose fracture damage was given by Wauer [147]. He surveyed 162 papers and also reviewed the non-rotating, crack structural elements which are relevant to the crack rotor problems. Extensive reviews of the dynamic response of cracked rotor systems were published by Dimarogonas [34]. He cited more than 350 papers and reviewed the vibration of stationary cracked beams and plates, the continuous crack beam and bar, crack identification in beams, vibration of cracked rotors, the closing crack and also described list of topics of interest for further research in vibration of cracked structures.

Sabnavis et al. [116] reviewed various crack detection techniques and diagnosis. They grouped the current research which is published after 1990 into three categories: vibration-based methods which can be sub-classified into signal-based and model-based methods, modal testing and changes in system modal characteristics, such as changes in mode shapes and system natural frequencies, response to specially applied excitation for crack detection. The last category is non-traditional methods such as neural networks, fuzzy logic, borescopes inspection and sophisticated signal processing techniques, e.g. wavelet and Wigner-Ville transforms. Kumar and Rastogi [74] cited more than 60 papers and surveyed various methods like Wavelet transform, FEM, non linear dynamics, Hilbert-Huang transform, and analysis of cracked rotor through various other techniques. More recent studies have been reviewed by Bachschmid et al. [10]. They noted that the breathing mechanism of cracks in rotating shafts is accurately investigated by means of 3D non-linear models. The behaviour is then modelled by means of a much simpler approximated approach, which allows also to calculate the stiffness variation of the cracked shaft.

Dimarogonas and Papadopoulos [35] reviewed the analytical method for the computation of the dynamic response of cracked Euler-Bernoulli beams. They modelled the cracked region as a local flexibility found with fracture mechanics method which is discussed in textbook by Dimarogonas [33]. Gasch [40] considered the non-linear mechanism of a closing crack with different flexibilities for the open and closed crack. He derived the equation of motion in the rotating and stationary coordinate systems. These equations were solved in an analog computer and the crack flexibility was measured by experiment. He also reported the sub-harmonic resonance at approximately half and one third of the bending critical speed of the rotor to be the prominent crack indicators. Mayes and Davis [89] performed a detailed analytical and experimental investigation for turbine shafts with cracks.

They derived a rough analytical estimation of the crack compliance based on the energy principle of Paris and measured it on a test rig. They obtained analytical solutions by considering an open crack, which leads to a shaft with dissimilar moments of inertia in two perpendicular directions, a problem with a known analytical solution. Grabowski [49] argued that in shafts of practical interest the shaft deflection due to its own weight is orders of magnitude greater than the vibration amplitude. Therefore he suggested that non-linearity is not affecting the shaft response since the crack opens and closes regularly with the rotation. Therefore the equations of motion can be considered linear with variable coefficients.

A very simple model of a flexible rotor is the so called de Laval rotor, sometimes also called Jeffcott rotor. This rotor system has historical meaning due to the fact that Jeffcott published the theory about this system in 1918, while de Laval investigated experimentally the self centering effect of the rotor already in 1883. [122], [112]. Today, the Jeffcott rotor is often used in order to explain the basic dynamic behaviour of a flexible shaft with a mass located in the shaft center. Dimarogonas and Papadopoulos [35] investigated the de Laval rotor with an open crack. They carried out analysis of cracked rotors neglecting the non-linear behaviour of the crack by assuming constant stiffness asymmetry. Furthermore, analytical solutions are obtained for the closing crack under the assumption of large static deflections. The computation of the local flexibility was based on the plane strain assumptions for the shaft and they used the stress intensity factors for the plane strip, since such factors are not available for the transverse crack on a rotating cylindrical shaft. Gasch [41] demonstrated that opening and closing of the crack during rotation is mainly due to shaft self-weight. He provided a simple but comprehensive survey of the stability behaviour of a rotating shaft with a transverse crack, and of the forced vibrations due to imbalance and to the crack. By utilizing a single parameter 'hinge' crack model, he assumed weight dominance and employed a perturbation method into his analysis. Mayes and Davies [89] analysed experimentally and theoretically the effects of a transverse crack on a rotor. They calculated the dynamic response of a cracked shaft and concluded that its non-linear response depends on the phase between unbalance and position of the crack. Mayes and Davies [90] introduced a model which is more practical for deep cracks than a hinged model and this model is called Mayes' steering function. They suggested sinusoidal stiffness variation to model the breathing in a more accurate if the crack opens and closes gradually due to gravity.

Jun et al. [66] derived the equations of motion for a simple rotor with a breathing crack based on LEFM, and the breathing crack model is further simplified to a switching crack model. By using the switching crack model, the conditions for crack opening and closing are derived. They proposed a response dependent breathing crack model. The model iteratively estimates the status of the crack closure using forces acting on the crack section, and thereby evaluated the stiffness of the cracked section. In this way, the model accounts for partial opening and closing of crack. Patel and Darpe [104] devoted nonlinear dynamics of the flexible cracked de Laval rotor on simple rigid supports. Their work examined the non-linear character of the cracked rotor in the subcritical speed range, using two well-known crack models; switching crack and breathing crack. They reported that unbalance phase, unbalance level, depth of crack and damping in the system have significant influence on the nonlinear vibration features of the cracked rotor with switching crack but have no influence on the rotor with breathing crack.

There are obviously many different approaches to the problem of cracked rotor dynamics studies the coupling between lateral, longitudinal, and torsional vibrations. Papadopoulos and Dimarogonas [101], [100], [103] extensively addressed the issue of coupling of vibrations due to a crack. They proposed the presence of either of bending, longitudinal or torsional mode natural frequency in the vibration spectra of the other modes as a potential indicator of a crack in the shaft. For this purpose they used harmonic sweeping excitation. The excitation, however, is given to non-rotating shaft. The other authors discussed some recent issues in coupled vibrations due to the crack are Chasalevris and Papadopoulos [18], Dado and Abuzeid [28], Sekhar [124], [123], Sekhar and Prabhu [126] and Sekhar and Prasad [129].

Ostachowicz and Krawczuk [98] presented the influence of transverse cracks on the coupled torsional and bending vibrations of a rotor. They developed the stiffness matrix for a beam element containing a single-sided open crack using finite element (FE) and fracture mechanics technique. Sekhar and Prabhu [127] used FE model for the cracked rotor with open crack and studied possibility of backward whirl and fluctuation of bending stresses due to crack. Darpe et al [30] presented accounts for coupling between longitudinal, lateral and torsional vibrations for a rotating cracked shaft using a response-dependent non-linear breathing crack model. By including the axial degree of freedom in their analysis, the stiffness matrix formulated is an extension of the one developed by Ostachowicz and Krawczuk [98]. They used a refined breathing crack model that accounts for partial opening and closing of crack through sign of SIF at the crack edge. In addition, they also proposed a concept of crack closure line to be able to study the flexibility variation with amount of crack opening. They assumed that the crack closure line is an imaginary line perpendicular to the crack edge.

Muszynska et al. [92] discussed rotor cross coupled lateral and torsional vibrations due to unbalance, as well as due to shaft asymmetry under a constant radial preload force. In their experiment, an asymmetric shaft was used to simulate the behaviour of a crack. Wu and Meagher [154] investigated the vibration response of cracked and asymmetric shafts. Non-dimensional analytical models of extended de Laval rotors are derived from Lagrange's equations taking into consideration the lateral/torsional vibration coupling mechanism induced by a breathing crack or a geometry asymmetry.

Many different approaches to model the behaviour of a crack in rotating structures have been proposed and published. In the first one, the crack is assumed to open and close fully following a square pulse function (Jun et al. [66], Patel and Darpe [104]). This behaviour is connected to a rather small crack, while a deep crack could be associated with a harmonic variation (Papadopoulos [99]). Another model is the regularly periodic breathing crack model for heavier rotors. This crack may gradually open and close during rotation. A common model of this behaviour can be found in Mayes and Davies [90], Gasch [41], Sinou and Lees [136], [137], where the opening and closing of the crack is described by a cosine function. The cosine function model assumes that the gravity force is much greater than the unbalance force. Furthermore, no direct relationship between the shaft stiffness and the material properties of the shaft exist.

Another approach is the breathing crack model using SIFs (Jun et al. [66], Papadopoulos and Dimarogonas [101], [103], Darpe et al. [30], Arem [7], Daoud et al. [29], Kisa et al. [73], Qian et al. [110]. In general, the equations of motion for a simple rotor with a breathing crack were derived using the local flexibility concept and the fracture mechanics approach for its computation (Dimarogonas and Papadopoulos [35]). A crack in a rotor is often simulated using the switching crack model, also known as hinge model, or the breathing crack model. In real cracked rotors, the stiffness variation is likely to be gradual, the assumption of abrupt stiffness switching is not appropriate. Hence, the switching crack model might adequately represent very shallow cracks and should be used with care to predict the vibration characteristics of the cracked rotor, particularly for deeper cracks, which can cause chaotic and quasi periodic vibrations [104]. In contrast, in the breathing crack model, the amount of open part of the crack continuously changes with the shaft rotation, thereby partial opening and closing of the crack are accounted.

Chondros et al. [23] developed a consistent continuous cracked beam theory. They used LEFM methods to model the crack as a continuous flexibility in the vicinity of the crack region investigating the displacement field. They also proposed that LEFM allows the development of a consistent cracked beam vibration theory without assumptions for the stress field. Penny and Friswell [106] used a simplistic model of the non-linear behaviour of a beam with a closing crack. Bachschmid and Tanzi [9] investigated the deflections of a beam with circular cross-section presenting a transverse crack of 50% depth caused by various loads. The characteristic breathing behaviour of the cracked area was also analysed and compared to that obtained with a rather simple one-dimensional model, which was compared to experimental results. They emphasized the differences between experimental results and computational predictions based on LEFM. Georgantzinos and Anifantis [47] studied the effect of the crack breathing mechanism on the time-variant flexibility due to the crack in a rotating shaft considering quasi-static approximation. The effect of friction is also considered in the cracked area. Portions of crack surfaces in contact are predicted, and direct and cross-coupled flexibility coefficients are calculated by applying energy principles. Kisa and Brandon [72] developed a FE model, the component mode synthesis method and the LEFM theory to compute the eigensystem for a cracked beam for different degree of closure. In addition, contact phenomena when a crack opens and closes during vibration are taken into account. Boubolas and Anifantis [15] developed a FE model in order to study the vibrational behaviour of a beam with non-propagating edge crack. The beam is discretized into FEs while the breathing crack behaviour is treated as a full frictional contact problem between the crack surfaces.

Since analytical SIFs for an edge crack in a rotating cylinder are not available, the shaft is considered to be a sum of elementary independent rectangular strips (Andrieux and Varé [5]) and no interaction between them is assumed to take place (Chasalevris and Papadopoulos [19]). The geometric functions used that describe the strain energy density are often not accurate enough, due to the fact that the crack passes from stress state caused by the vertical moment to that of horizontal moment. Then, the compliance is obtained by integrating along the crack tip. If the crack depth exceeds the radius of the shaft, then the elements of the compliance matrix present a divergence. This is due to a singularity that the strain energy release rate method has near the edges of the crack tip, giving thus the infinite values. It was reported by Papadopoulos [99], that divergence does not reflect reality. Furthermore, the relative crack depth or crack depth ratio is the only parameter of the crack geometry for governing the SIF in the reference cited above. During crack growth, the relative crack depth is not enough to describe the crack geometry. Shih and Chen [131], [132], develop a two parameter relationship between SIF and crack geometry which accounts for crack aspect ratio (ratio between crack depth and crack length) and relative crack depth. Two issues related to the critical speed of a cracked rotor have been emphasized [21]. One of these is the stability of a cracked rotor near its critical speed ([8], [62], [58], [128], [125], [135], [102], [20], and [157]). The other is the transient response when the rotor passes through its critical speed ([108], [31], [128]).

1.2.2 Cohesive zone model

The CZM describes material failure on a phenomenological basis (i.e. without considering the material microstructure). The advantage of this model when compared to LEFM is that the parameters of the respective models depend only on the material and not on the geometry. This concept guarantees transferability from specimen to structure over a wide range of geometries. Furthermore, CZMs are surfaces of discontinuities where displacements are allowed to jump. A specific constitutive law relating the displacement jumps and proper traction defines the CZM. Within the cohesive zone approach crack nucleation, propagation, and arrest are a natural outcome of the theory. The latter is in contrast to the traditional approach of fracture mechanics where stress analysis is separated from a description of the actual process of material failure. The origin of the CZM can be traced back to the strip yield model proposed by Dugdale [37] and Barenblatt [14] in which the narrow zone of localized deformation ahead of the crack tip was substituted by cohesive traction between the bounding surfaces. Dugdale used the concept of a simple fracture process zone, where the closing stress is assumed to be constant, to model the plane stress plastic zone at a tip of a crack as a fictitious extension to the real crack using the condition of smooth closure to find the extent of the plastic zone. Barenblatt was the first to formulate fracture mechanics in terms of a fracture process zone. Considering an elastic-brittle material, he proposed that there was an inner zone to the crack where the atomic cohesive forces, dependent on the opening, were important. The difference between Barenblatt and Dugdale was that the former was modelling the interatomic forces at the crack tip and the latter using the concept of a fracture process zone to model plastic deformation.

The fracture process zone model into two main versions is classified by Cotterell [26], namely the fictitious crack model pioneered by Hillerborg and the crack band model proposed by Bažant. He presented that modelling the fracture process zone with the classic one parameter models of elastic and plastic fracture have been very successful, but they have two major limitations: they can only be used if there is an initial crack, and the fracture process zone must be small compared with the dimensions of the specimen. Classic elasto-plastic models have a third limitation, only small crack growth can be modelled. Elices et al. [38] reviewed the cohesive process zone model, a general model which can deal with the nonlinear zone ahead of the crack tip due to plasticity or microcracking present in many materials. They described that LEFM has proven a useful tool for solving fracture problems provided a crack like notch or flaw exists in the body and the nonlinear zone ahead of the crack tip is negligible. This is not always the case, and for ductile metals, the size of the nonlinear zone due to plasticity or microcracking is not negligible in comparison with other dimensions of the cracked geometry. Moreover, even for brittle materials, where the process zone can be lumped into a single point, the presence of an initial crack is needed for LEFM to be applicable. This means that bodies with blunt notches but no cracks cannot be analysed using LEFM.

The constitutive behaviour which causes the cohesive elements to open and eventually to fail is described by the so called TSL. It relates the traction vector to the displacement jump across the interface. Various functions have been proposed for the shape of the TSL. Common to all shapes is that they contain two model parameters, namely the maximum traction sustainable by the element and a critical separation at which the element finally fails, that was reported by Hutchinson and Evans [59]. The energy dissipated by the element until total failure is derived as the integral of the TSL. Volokh [146] examined the fracture prediction done by four different CZMs considering a block-peel test. He concluded that the role of the shape of the CZM is essential in describing the fracture process. Jin and Sun [63] investigated a comparison of CZM and fracture mechanics based on near tip stress field for mode III crack. A review of the theory concerning CZM applications to various materials was made by Siegmund and Brocks [133]. CZMs have been used to simulate the fracture process in a number of material systems including metallic materials, ceramic materials, concrete, polymer and fiber reinforced plastic composites. They have been used to simulate fracture under static, dynamic and cyclic loading conditions. Xu and Needleman [155] used the CZM to study the void nucleation at the interface of particle and matrix material, Tvergaard and Hutchinson [145] used a trapezoidal shape of the TSL to calculate the crack growth resistance in elastic-plastic materials.

Shet and Chandra [130] analyzed energy balance when using CZM to simulate fracture processes. Li and Chandra [79] analyzed crack initiation and crack growth resistance in elastic plastic materials, dominated by crack tip plasticity with the crack modelled as a CZM. Geubelle and Baylor [48] utilized a bilinear CZM to simulate spontaneous initiation and propagation of transverse matrix cracks and delamination fronts in thin composites plates subjected to low-velocity impact. Turon et al. [144], [143] used the bilinear CZM to determine the constitutive parameters for the simulation of progressive delamination (or interfacial cracking between composite layers). Their procedure accounts for the size of a cohesive FE and the length of the cohesive zone to ensure the correct dissipation of energy. Alvano and Sacco [4] proposed a new method to combine interface damage and friction in a CZM. Cornec et al. [25] proposed a method for practical application of the CZM with emphasis on metallic material. The method used consists of a defined shape of the TSL, determination of the parameters for the TSL, and suggestions for FE analysis. In addition, they implemented the CZM using the programming language FORTRAN as a user defined element within the FE code ABAQUS.

The crack growth in elastic-plastic materials that exhibit ductile fracture, i.e. induced by void growth and void coalescence was reported by Scheider [118]. There exist two approaches to the modelling of material failure Gurson-Tvergaard-Needleman (GTN) model and CZM with constant parameters (numerically fitted to the experiments) and triaxiality dependent parameters (determined from GTN unit cell). However, the TSL depends on the stress state, which can be characterised by the triaxiality, which is the hydrostatic stress divided by the von Mises equivalent stress. This issue was first investigated by Siegmund and Brocks [134]. The approach was extended to simulation of dynamic ductile crack growth by Anvari et al. [6]. Banerjee and Manivasagam [13] proposed a versatile CZM to predict ductile fracture at different states of stress. The formulation developed for mode-I plane strain accounts explicitly for triaxiality of the stress-state by using basic elastic-plastic constitutive relations combined with two model parameters, which are independent of the stress-state. The importance of the CZM concept for describing fracture in a wide range of engineering materials has been recognized over the past few decades. However, its proper numerical implementation has caused problems. Essentially, the CZM is a discrete model and cannot be implemented readily in standard, continuum-based FEM. Crack paths are normally not known in advance. To partly circumvent this difficulty, proposals have been made to insert interface elements between all continuum elements, to carry out remeshing procedures or to use mesh-free methods. A consistent extension to dynamic problems is possible.

1.3 Objectives

The purpose of this dissertation is to address the application of CZM of a cracked rotating shaft and to analyze the dynamic behaviour of a shaft with transverse breathing crack.

The present study is organized to tackle the following questions and issues.

- What are advantages from using CZM in dynamic problems such as vibrations of a rotating shaft instead of fracture mechanics approach?

- It is known that stiffness of a cracked rotating shaft varies with time due to breathing mechanism. Can the stiffness variation of a cracked shaft be modelled using CZM?, and how is the result compared to the breathing crack model based on LEFM?

- How is the stability of a rotor system affected by a transverse breathing crack?

- How can CZM in conjunction with FEM be implemented for a cracked rotor model?

- Which parameters have significant influence on the breathing mechanism during rotation of a shaft?

In Figure 1.2, a flow chart of the conducted research is presented. Contributions of this research are expected to provide progress in modeling of cracked rotor and to provide more accurate and meaningful results.

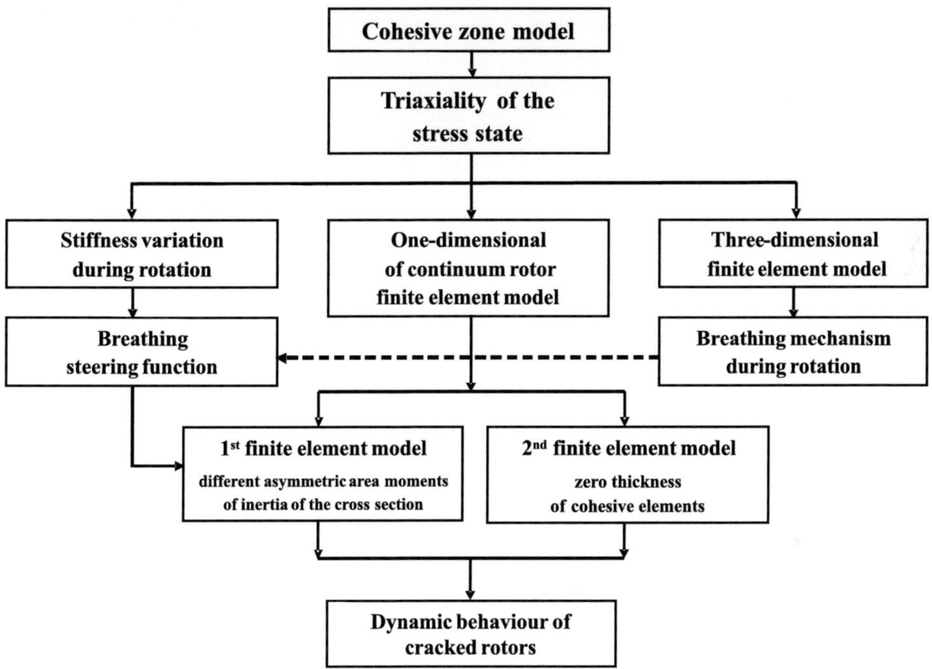

Figure 1.2: Research design

1.4 Structure of the dissertation

The dissertation is organized into seven chapters, and structured in a way to meet the objectives of the study. In particular:

In Chapter 1, motivation of research on the application of the CZM to the analysis of a rotor with a transverse crack is presented. Literature on dynamics of cracked rotor and CZM concepts is also reviewed. Objectives of the research including research design and method and structure of the dissertation are formulated systematically.

Chapter 2 introduces the definitions and theorems of the fracture process zone and CZM such as basis of the models, location of crack tip, then kinematic and constitutive relation of CZMs. The FE implementation including one cohesive element test, stress distribution on a cracked plate, simulation of delamination and energy balance concept are presented. This chapter also deals with the triaxiality dependent cohesive zone in ductile materials and emphasizes formulation of triaxiality dependent model and TSL for ductile materials. In the last section, an implementation of the CZM for a cracked shaft under rotating load is introduced. This chapter provides a theoretical background for the dissertation.

In Chapter 3, some different approaches of breathing steering function during rotation of shaft are discussed. Several breathing crack shape models reported in literature are also presented. The simple cracked rotor model or a de Laval rotor is extended to account for the changes of stiffness during rotation. First, shaft stiffness variation due to the breathing crack during rotation based on LEFM is reviewed. Then, the CZM is applied to estimate the shaft stiffness variation on the cracked rotor. Breathing crack shape is modelled by a parabolic line for deep crack and by a straight line for a shallow crack.

The instability due to parametric excitation of simple cracked rotor models treated in previous chapter are investigated in Chapter 4. Perturbation method to obtain the boundaries of stability regions is applied. In this chapter is also shown that some small damping in the system is very helpful to guarantee stability.

Chapter 5 presents FE modelling of the cracked shaft based on equivalent beam using CZM. Two FE models one dimensional continuum rotor are proposed. The first model is based on different asymmetric area moments of inertia of the cross section due to the crack. The other model is based on the TSL using one cohesive element with zero thickness which is placed between the continuum elements. Chapter 5 also deals with the implementation of CZM in FE to predict and to analyse the dynamic behaviour of cracked rotor. The results are compared with the classical analytic method; Timoshenko beam theory for open cracked shaft without disk. Dunkerley's equation and Rayleigh's method are also used to estimate lower and upper bounds of the fundamental natural frequency from CZM for a cracked shaft with disk.

In Chapter 6, the breathing mechanism of a simple cracked shaft on rigid supports has been investigated. An integrated simulation process of FE and multi-body simulation (MBS) is employed. At the first step, an elastic cracked shaft with various relative crack depths is modelled by FE. The analysis deals with the natural frequencies and mode shapes of the cracked rotor and the uncracked one. Furthermore, breathing crack under rotating load (non-rotating shaft) is also investigated. At the second step, the FE model of an elastic cracked shafts are exported into MBS in order to analyze the dynamic loads, due to the crack, unbalance and inertia force acting during rotation at different rotating speeds. The effect of orientation angle of the unbalance mass on the cracked rotor has also been demonstrated. Finally, the vibration responses in the centroid of the shaft obtained from MBS have been exported into FE to observe the breathing mechanism. In the last section of Chapter 6, the proposed model for breathing crack in Chapter 3 is validated by comparing to the breathing mechanism obtained by an integrated simulation process of FE and MBS.

Chapter 7 concludes the dissertation by advancing a framework on application of CZMs for the investigation of the dynamic behaviour of a transverse cracked rotor and outlining the contribution to theory, limitations, directions for future research, as well as some implications of the findings for practice.

In Figure 1.3, the organization of core chapters is presented along with the interaction between chapters and their subsections.

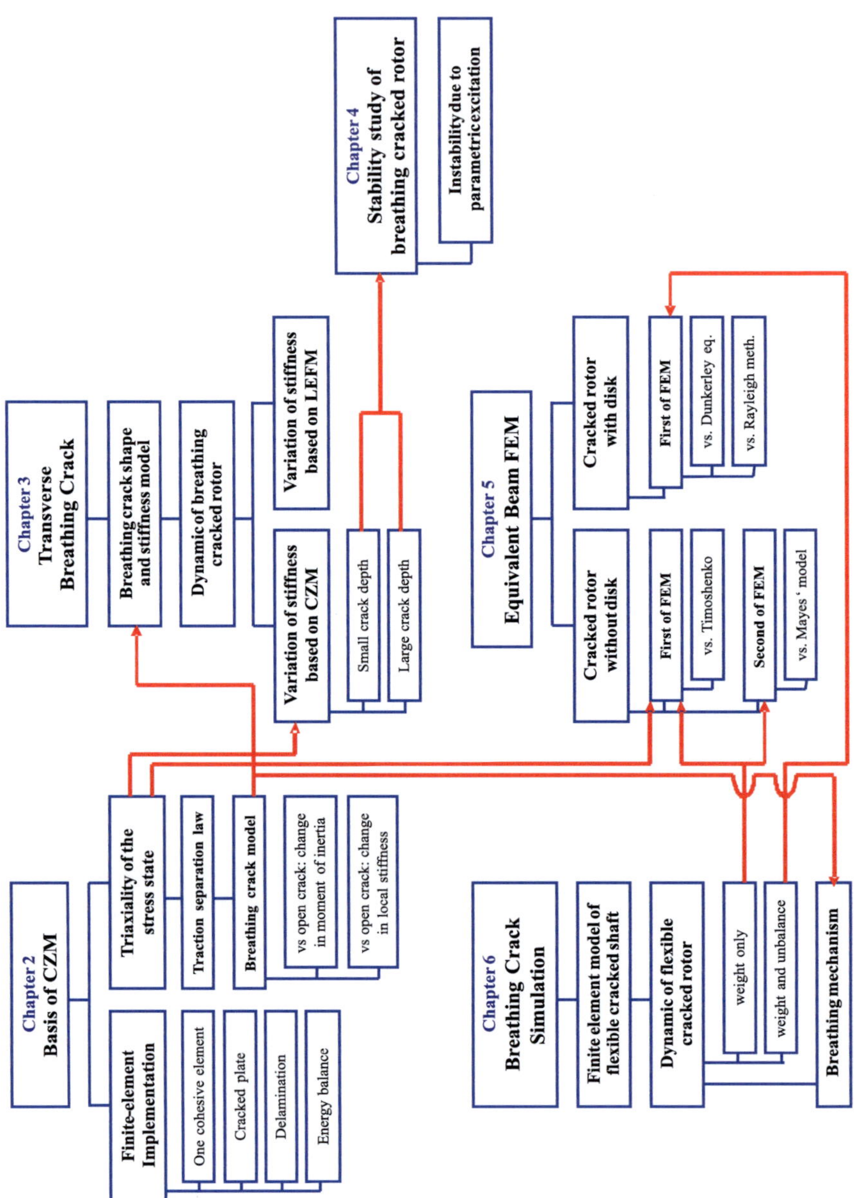

Figure 1.3: Structure of the dissertation and corresponding chapters

2 Cohesive zone models

This chapter presents some basic aspects of fracture process zone. Three types of approaches to model the fracture process zone are introduced namely; fracture mechanics, cohesive zone model (CZM) and damage mechanics. A large section of this chapter is devoted to the values of the elastic limit and failure threshold, as well as an overview of CZM for brittle and ductile materials. Fundamentals of CZM and location of crack tip, further, the mechanical behavior of CZM, basic of one dimensional kinematics and constitutive relation of CZM are introduced. This chapter aims at the implementation of CZM in modeling of a cracked rotor with a breathing crack, and it is essential to understand the basics and fundamentals of CZM. The CZM using TSL is implemented in the FE modeling. In order to test the accuracy of FE programming, firstly a simulation of one cohesive element is investigated. Another implementation, stress distribution on a cracked plate based on LEFM, the Irwin's model, the Dugdale's model and the CZM are explored. These simulations are made to show that the presence of cohesive elements in a zone ahead of the crack tip has stresses lowering due to the softening of cohesive elements. Simulation of delamination in the double cantilever beam (DCB) and the mixed-mode bending (MMB) test are included for investigating influence of interface stiffness parameter. Other section of this chapter examines how the external work flows as recoverable elastic strain energy, dissipated plastic energy and cohesive energy. The CZM allows the energy to flow into the fracture process zone.

This chapter also includes a section on triaxiality dependent CZM for ductile materials and discusses TSLs for ductile materials. In particular, implementation of the CZM for a cracked shaft is introduced using the FE with a rotating loading. Although in this chapter the dynamic behavior of a cracked rotor is not yet discussed, the breathing mechanism of non rotating cracked shaft under rotating loading is demonstrated and compared to the results in literature. Two approaches of the open crack model based on change in second moment of area and change in local stiffness are introduced in order to compare with the breathing crack mechanism of a non rotating shaft based on CZM.

2.1 Fracture process zone

Fracture mechanics in general is used to understand and predict structural failure due to the presence of macroscopic cracks, which could not be captured by classical theories of elasticity and plasticity. The theory of fracture mechanics is strongly based on the theory of elasticity and many empirical solutions to simple problems have been derived for general use. General application of empirical solutions requires simplification of the actual problem to match the empirical basis, often leading to uncertainty in the accuracy. The crack tip as addressed in fracture mechanics is a mathematical idealisation, but in reality, a region of material degradation exists ahead of a micro-crack, the so-called fracture process zone, where finally new surfaces are created. Identifying the micro-mechanisms occurring in the fracture process zone is crucial for a fundamental understanding and modelling of macroscopic fracture.

The fracture process zone can be modelled using the following three approaches [16];

1. **Fracture Mechanics.** LEFM is applied when the fracture zone is surrounded by an elastic region that can be characterized by stress intensity factors in case of linear elasticity or J-integrals in case of nonlinear elasticity. The process zone is a point, where stresses and strain may become singular. Fracture mechanics parameters like J-integral, or crack-tip opening displacement (CTOD) are introduced, which under certain conditions dominate the crack-tip field. No splitting of dissipation into global plasticity and local separation is possible. Numerical simulation of crack extension is based on node release following an R-Curve, $J(\Delta a)$ or $\delta(\Delta a)$, where Δa is the macroscopic crack extension. Fracture mechanics has been used to model crack problems on the assumption of the existence of initial defects or cracks and cannot be applied directly without initial defects or cracks.

2. **Cohesive Zone Model.** Separation of surfaces is admitted, which is governed by a special cohesive law. The process zone is now a two-dimensional surface, and parameters like cohesive strength and separation energy characterise the local material failure. Dissipation can be split into contributions by global plasticity and local material separation. Respective cohesive elements are introduced at the interfaces of continuum elements. Whereas these interface elements obey a special TSL the material outside is described by conventional constitutive equations of elasto-plasticity. These models relate traction to the relative displacement or separation at an interface, where a crack may occur. Crack or damage initiation is related to an interfacial strength, i.e. the maximum traction on the traction-separation curve. When the area under this curve is equal to critical fracture energy, the traction is reduced to zero and complete crack surfaces are formed.

3. **Damage Mechanics.** A softening constitutive behaviour accounting for damage is quantified by internal variables like porosity, micro-crack density, etc. The process zone is a volume. Damage models describe the evolution of degradation phenomena on the microscale from initial (undamaged or predamaged) state up to creation of a crack on the mesoscale (material element). From a physical point of view, damage is always related to plastic or irreversible strains and more generally to a strain dissipation either on the mesoscale, the scale of the representative volume element, or on the microscale, the scale of the discontinuities. Lemaitre and Desmorat [78] classified the level of damage as follows: in the first case (mesolevel), the damage is called ductile damage if nucleation and growth of cavities in a mesofield of plastic strains under static loading occurs. It is called creep damage when it occurs at elevated temperature and is represented by intergranular decohesions in metals. It

is called low cycle fatigue damage when it occurs under repeated high level loadings, inducing mesoplasticity. In the second case (microlevel), it is called brittle fracture, or quasi-brittle damage, when the loading is monotonic. It is called high cycle fatigue damage when the loading is a large number of repeated cycles. In all cases, there are volume defects such as microcavities, or surface defects such as microcracks.

Fracture mechanics studies are usually carried out under several idealized conditions, as in case of LEFM or in case of small scale yielding. Details of the local crack tip fields are uniquely characterized by a single macroscopic parameter such as the SIFs or corresponding energy release rates. These global parameters are related to the corresponding material parameters (typically the fracture toughness or critical energy release rates) that determine the critical conditions of initiation of crack growth. When the crack tip experiences plastic yielding, the above concepts based purely on the theory of elasticity are not valid and have led to the introduction of a path independent J-Integral, which is strictly valid for a nonlinear elastic material. The property of path independence is lost if the energy near the crack tip region is converted into significant inelastic energy due to plasticity or when the material locally unloads during the propagation process. Fracture mechanics analysis assumes the existence of an infinitely sharp crack leading to the singular crack tip fields. However, in real materials neither the sharpness of the crack nor the stress levels near the crack tip region can be infinite.

Recently, the CZM has emerged as a powerful tool for investigating the fracture processes in materials and structures. The origin of cohesive zone concept can be traced back to the strip yield model by Dugdale [37] and Barenblatt [14] in which the narrow zone of localized deformation ahead of the crack tip was substituted by cohesive traction between the bounding surface. Dugdale and Barenblatt proposed the concept of CZM as an alternative approach to the singularity driven by fracture mechanics approach. In the CZM model, a finite stress distribution in the vicinity of the crack tip involving interaction between the crack faces is considered instead of just the singular crack tip. CZM has evolved as a preferred method to analyze fracture problems in material systems not only because it avoids the singularity but also because it can be easily implemented into an existing FE code via an interface element. The CZM represents a narrow band of localized deformation and is idealized as a pair of surfaces on which cohesive tractions act. The cohesive traction is defined as a function of the separation displacement in the form of TSL or cohesive law. The Traction-separation relations for the interface are such that with increasing interfacial separation, the traction across the interface reaches a maximum, then decreases and eventually vanishes, permitting a complete decohesion. There is common belief that CZM can be described by two independent parameters (Hutchinson and Evans [59], Elices *et al.* [38]). These parameters may be two of the three parameters, namely the cohesive energy, and either of the cohesive strength (traction) or the separation (relative displacement). In general, cohesive energy is obtained from experiments and is believed to be equivalent to the work of fracture; the latter quantity identified as, for example critical strain energy release rate or cohesive energy G_{IC} or critical J-energy or fracture energy J_{IC}.

2.2 Ductile and brittle fracture

Elastic material is defined as the material that returns to its original configuration when deforming forces removed. Although no real material is perfectly elastic, a brittle material is simply one that behaves elastically up until the point where it fractures. In contrary to an elastic material, a plastic material will not return to its previous configuration once deforming forces have been removed. As stated before, real materials do not behave perfectly elastic. Real materials can be deformed only to a limited extent before they will no longer return to their previous configuration. This limit is known as the elastic limit or yield limit. When the elastic limit has been exceeded, the material enters a plastic regime and begins to experience plastic flow. Eventually, at the failure threshold, it fractures.

The terms brittle and ductile relate to the relative values of the elastic limit and failure threshold. If the failure threshold nearly coincides with the elastic limit, then the material will experience only negligible plastic deformation before fracture. The term brittle refers to such a material. In contrast, for a ductile material the failure threshold is significantly larger than the elastic limit so that as the material deforms it experiences an elastic regime, followed a plastic regime, and then finally fracture.

The significance of the distinction between ductile and brittle materials arises because elastic deformation stores energy whereas plastic deformation dissipates it. When a brittle material is deformed to its failure threshold, the majority of the energy used to deform it is stored as elastic potential. When fracture occurs, the energy is released and it tends to drive the fracture crack into the material. Thus, even though a large or small force may be required to start a crack in a brittle material (depending on its toughness), once the crack is started only a small amount of energy is required to push it further. In contrary to the ductile material requires significantly more work to propagate a crack because energy is being absorbed by plastic deformation. As a result, brittle materials tend to break at or shortly past their elastic limit (almost no plastic deformation), while ductile materials deform at stress levels beyond their elastic limit. In general the underlying causes of plasticity are fairly complicated and they give rise to a number of phenomena. For example, the energy absorbed by plastic deformation does not simply vanish and it may result in effects such as fatigue weakening.

Ramberg and Osgood [111] introduced an equation to describe the non linear relationship between stress and strain in materials near their yield limit, which can show whether the materials are ductile or brittle. It is especially useful for metals that harden with plastic deformation (strain hardening), showing a smooth elastic-plastic transition, and it has the form

$$\varepsilon = \frac{\sigma}{E} + K \left(\frac{\sigma}{E} \right)^{n} \tag{2.1}$$

where ε is strain, σ is stress and E is modulus of elasticity.

The first term on the right side σ/E, is equal to the elastic part of the strain, while the second term $K(\sigma/E)^n$ accounts for the plastic part. Parameters K and n are constants that describe the hardening behavior of the material and depend on the material being considered. Introducing the yield strength of the material σ_Y, and defining a new parameter α_m related to K as $\alpha_m = K(\sigma_Y/E)^{n-1}$, it is convenient to rewrite the last term of the right side as follows

$$K\left(\frac{\sigma}{E}\right)^n = \alpha_m \frac{\sigma_Y}{E}\left(\frac{\sigma}{\sigma_Y}\right)^n \tag{2.2}$$

Substituting Eq.(2.2) into Eq.(2.1), the Ramberg-Osgood equation can be written as

$$\varepsilon = \frac{\sigma}{E} + \alpha_m \frac{\sigma_Y}{E}\left(\frac{\sigma}{\sigma_Y}\right)^n \tag{2.3}$$

2.3 Cohesive zone model

2.3.1 Basis of the models

In the CZM, fracture nucleates as discontinuity surface able to transmit tensile load before opening above a given displacement. Formation and extension of this surface require that the maximum principal stress reaches a given value, namely the cohesive strength of the material. When this occurs, the surface initiates or grows perpendicularly to the direction of the maximum principal stress. The two faces of the surface exert on each other equal and opposite tensile stresses (cohesive stresses) whose value is a unique function $f(\delta)$ of the separation δ between the faces. When the separation reaches another given value (the critical separation, δ_c or δ_{sep}), the cohesive stress vanishes and fracture takes place. Fracture consists of the initiation and propagation of a crack produced by the opening and advance of the cohesive zone (the zone where the cohesive stresses act) ahead of the crack tip as shown in Figure 2.1.

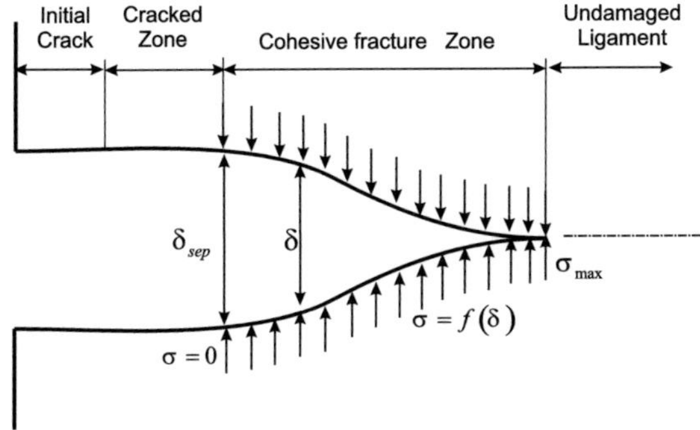

Figure 2.1: Fracture process zone model

The fracture behaviour of each material is described by the cohesive traction as a function $\sigma_n = \sigma_{max} f(\delta)$, where σ_{max} is the peak value of traction (Figure 2.2). Besides the cohesive strength and the critical separation, a key value of this curve is the cohesive energy G_{IC} (for mode I) the area enclosed under the curve. The mechanical behaviour of the bulk material is independent of the softening function and can be described by any constitutive equation.

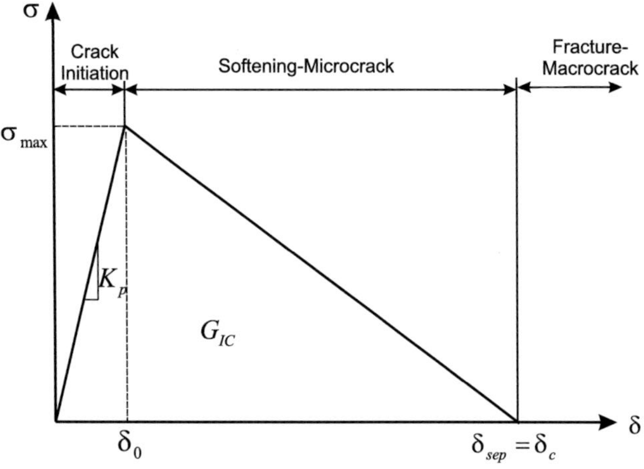

Figure 2.2: Traction separation law for brittle materials

2.3.2 Location of crack tip

The fracture process in CZM is shown in Figure 2.3, where the fracture process zone can be defined as the region within the separating surfaces where the surface traction values are nonzero. This also implies that processes occurring within the process zone are accounted for only through the traction-separation relations. In fracture mechanics, the crack growth problem is identified as a moving boundary value problem in which the primary unknown is usually the trajectory of a single point referred to as crack tip. CZM represents a zone or a region where material separates, the location of a crack tip within the cohesive process zone cannot be uniquely identified. There are a number of traction-separation forms available in literature [130], [79], [146]. In this work, the concept of identifying the location of the crack tip, by Shet and Chandra [130] is adopted. They proposed to assume the location of a cohesive crack tip at point (δ_0, σ_{max}) in order to interpret the numerical result. There seems to be no standard way of identifying the tip of the crack from traction and separation curve.

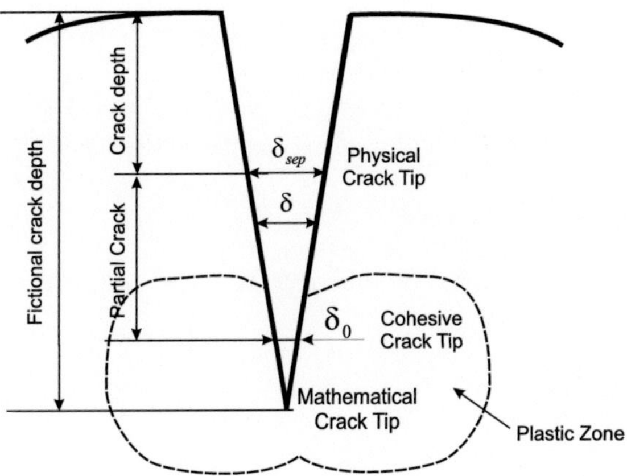

Figure 2.3: Fracture process in cohesive zone model

2.3.3 Kinematics and constitutive relation of cohesive zone models

The displacement jump across the interface of the material discontinuity required in the constitutive model $[\![u_i]\!]$, can be obtained as a function of the displacement of the points located on the top and bottom side of the interface u_i^+ and u_i^-, respectively

$$[\![u_i]\!] = u_i^+ - u_i^- \tag{2.4}$$

where $u_i^\pm$ are the displacements with respect to the fixed Cartesian coordinate system. A co-rotational formulation is used in order to express the components of the displacement jumps with respect to the deformed interface. The coordinate $\bar{x}_i$ of the deformed interface can be written as

$$\bar{x}_i = X_i + \frac{1}{2}\left(u_i^+ - u_i^-\right) \tag{2.5}$$

where X_i are the coordinates of the undeformed interface. The components of the displacement jump tensor in the local coordinate system on the deformed interface, δ_m are expressed in terms of the displacement field in global coordinates

$$\delta_m = \Theta_{mi}[\![u_i]\!] \tag{2.6}$$

where Θ_{mi} is the rotation tensor.

The constitutive operator of the crack interface, D_{ij} relates the element tractions σ_j to the displacement jumps δ_i

$$\sigma_j = D_{ij}\delta_i \tag{2.7}$$

This constitutive damage model was previously proposed by Turon *et al.* [142].

The free energy density per unit surface of the interface ψ is defined as

$$\psi\left(\delta, \mathrm{d}\right) = \left(1 - \mathrm{d}\right)\psi_0\left(\delta\right) \tag{2.8}$$

where d is scalar damage variable and $\psi_0(\delta)$ is the free energy per unit surface defined as

$$\psi_0\left(\delta\right) = \frac{1}{2}\delta_i D_{ij}^0 \delta_j \qquad i = 1,3; \quad j = 1,3 \tag{2.9}$$

The constitutive equation obtained by differentiating the free energy with respect to the displacement jumps is given as

$$\sigma_i = \frac{\partial \psi}{\partial \delta_i} = \left(1 - \mathrm{d}\right) D_{ij}^0 \delta_j \tag{2.10}$$

where D_{ij}^0 is the undamaged stiffness tensor and defined as

$$D_{ij}^0 = K_p \delta_{ij} \tag{2.11}$$

and K_p is a penalty stiffness.

2.4 Finite element implementation

In a FE model using the CZM approach, the complete material description is separated into fracture properties captured by the constitutive model of the cohesive surface, and the properties of the bulk material are captured by the continuum regions. The model shown in Figure 2.4 is implemented into a commercial FE code ABAQUS with its user element UEL subroutine coded in FORTRAN.

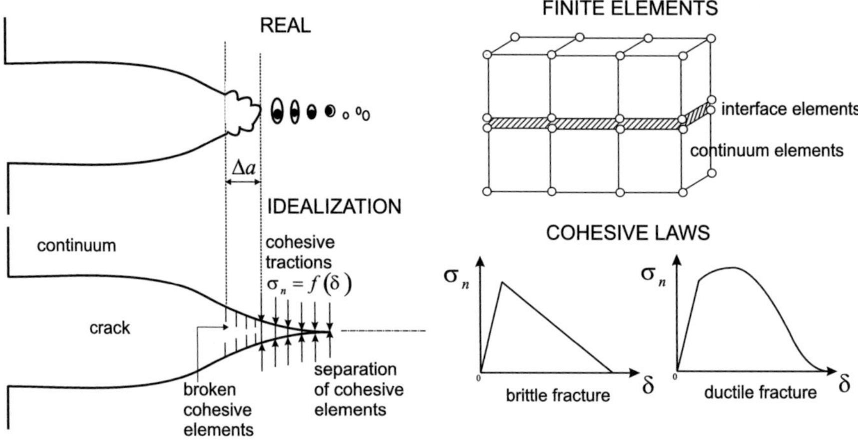

Figure 2.4: Representation of the fracture process using CZM in FE model

2.4.1 One cohesive element test

To examine the accuracy of finite element programming, first one cohesive element and one continuum element are used. The FE model shown in Figure 2.5 is composed of one 4-node plane strain element connected by a 4-node cohesive element representing the interface. The material properties used in the model were modulus of elasticity E, critical strain energy release rate G_{IC}, cohesive strength σ_{max} and penalty stiffness K_p.

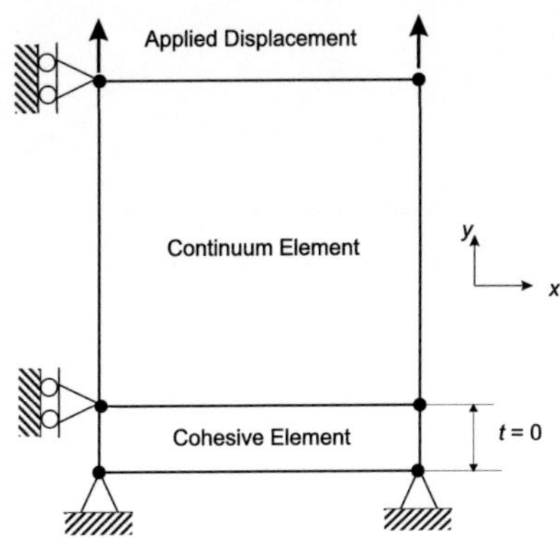

Figure 2.5: Simulation of one cohesive element

The load is applied in one step only in the normal direction (y-axis). Both traction and separation increased, till the traction reaches the peak value (σ_{max}) at a characteristic separation δ_0, as shown in Figure 2.6. Further, separation increases and traction decreases, until it vanishes completely, implying the creation of two new traction-free surfaces, i.e. crack growth. The energy required to create new surfaces in the area under the traction-separation curve, namely the cohesive energy is given as [97]

$$G_{IC} = \frac{1}{2}\sigma_{max}\delta_{sep} \tag{2.12}$$

$$= \frac{1}{2}K_p\,\delta_0\,\delta_{sep} \tag{2.13}$$

The material separation and the damage of the structure are described by interface elements at the boundaries of the undamaged continuum elements. Thus, the mechanical behaviour of the material is split in two parts, the continuum with any elastic-plastic constitutive law and the CZM specifying material damage and separation. For opening crack propagation mode or mode I, the function $\sigma(\delta)$ using bilinear TSL is described by

$$\sigma = \begin{cases} K_p\delta & \text{for} & 0 < \delta < \delta_0, \\ K_p\left(1 - \d\right)\delta & \text{for} & \delta_0 < \delta < \delta_{max}, \\ 0 & \text{for} & \delta \geq \delta_{max}. \end{cases} \tag{2.14}$$

where $\d$ is scalar damage of the material.

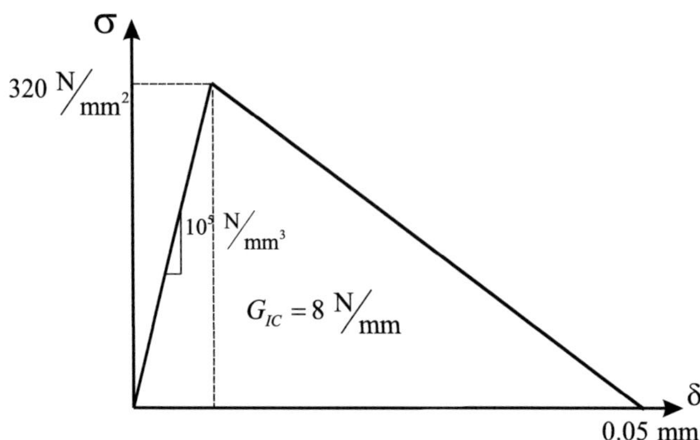

Figure 2.6: Bilinear traction separation law written in subroutine

The FE results using ABAQUS are the traction-separation curves (σ [N/mm²] vs. δ [mm]) and the area G_{IC} [N/mm] are shown in Figure 2.7. The results show that, when the maximum separation reached, the traction vanishes and the continuum element initially connected by the cohesive element is disconnected. The area of the traction-separation curve represents the cohesive energy as shown in Figure 2.6.

Figure 2.7a-c show the variation of cohesive strength σ_{max}. As can be seen, the maximal traction due to the load reaches until the peak value σ_{max} then decreases. All of the traction-separation curves in Figure 2.7a-c have the same areas as Figure 2.6. Figure 2.7d-f represent the different values of cohesive energy G_{IC}. If the cohesive energy increases, the area of the traction-separation curve does too. Variation of penalty stiffness K_p is depicted in Figure 2.7g-i (detail about penalty stiffness influence will be discussed in the next section). Figure 2.7j-l display the various values of applied displacement or separation δ while the other parameters are constant. In case applied displacement is greater than maximum separation δ_c, traction vanishes and no traction again until applied displacement reached.

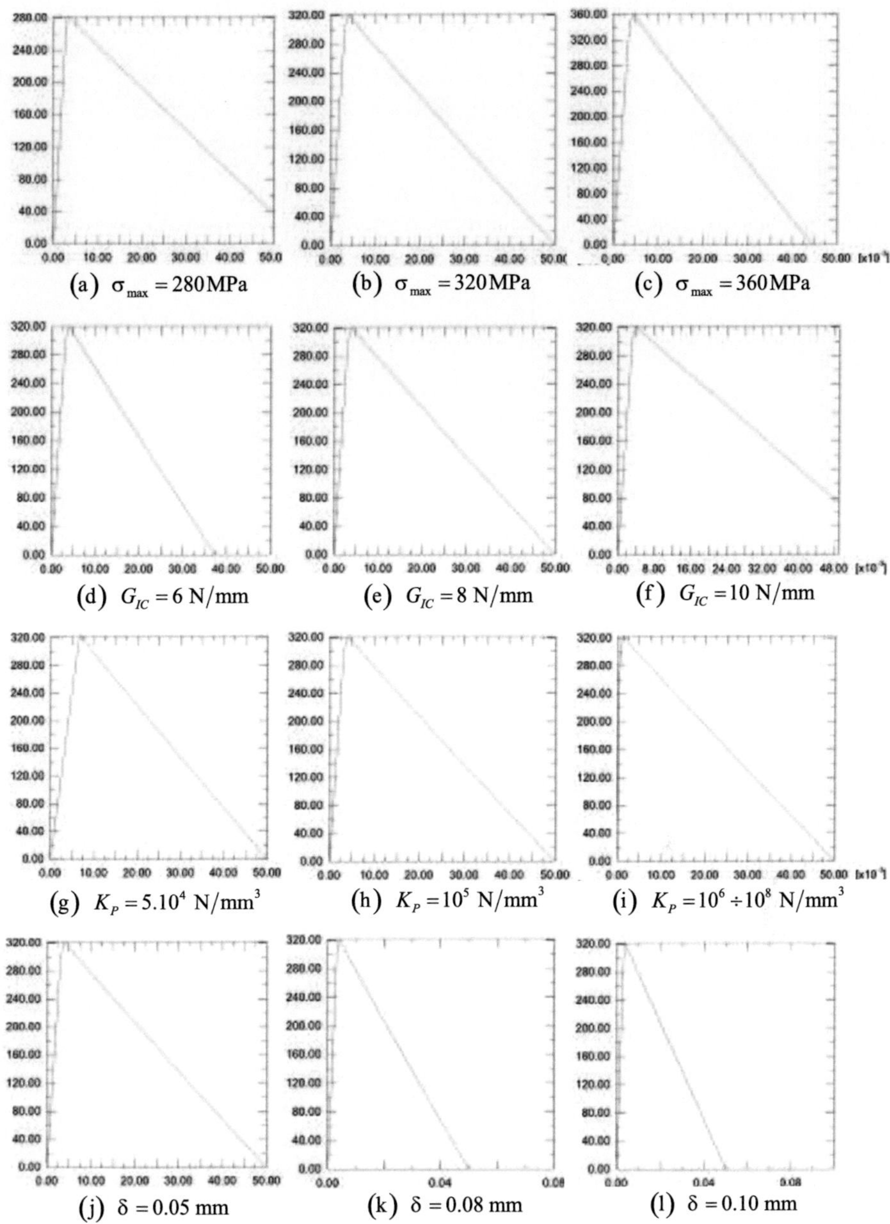

Figure 2.7: Finite element results of one cohesive element for different values of parameters, σ_{max}, G_{IC}, K_p and δ

2.4.2 Stress distribution on a cracked plate

In a second implementation, a thin steel plate of width $b = 1$ m containing length of edge crack of 0.2 m is subjected to a stress $\sigma = 97.6$ MPa normal to the crack plane as shown in Figure 2.8. The yield stress and the critical SIF are assumed to be 250 MPa and 66 MPa$\sqrt{\text{m}}$, respectively.

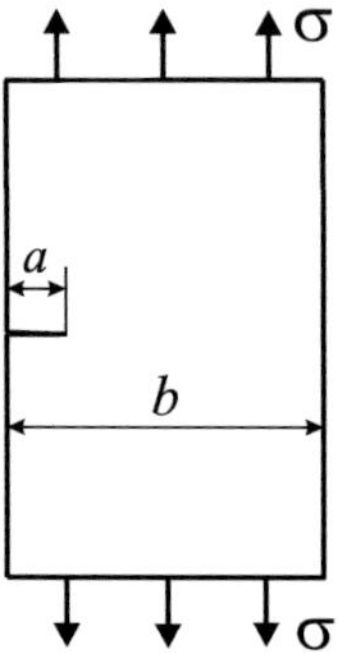

Figure 2.8: Thin steel plate contains an edge crack subjected to normal stress

2.4.2.1 Linear elastic fracture mechanics (LEFM)

In design, we would like to know the applied stress that causes fracture. This leads to an approach to LEFM that uses the SIF as the describing parameter. The SIF K_I, represents the amplitude of the crack tip stress singularity and is dependent on the body geometry, crack size, load level, and loading configuration.

The SIF for a cracked plate with crack length a for an applied stress σ is given by [45]

$$K_I = \sigma\sqrt{\pi a} \tag{2.15}$$

In a region near the crack tip, the stress is concentrated at the tip of a crack,

$$\sigma_y = \frac{\sigma\sqrt{\pi a}}{\sqrt{2\pi x}} = \frac{K_I}{\sqrt{2\pi x}} \tag{2.16}$$

where x is the distance ahead of the crack as shown in Figure 2.9.

For a single edge cracked plate under uniform tension, there is a geometry factor $F(a/b)$, which has to be taken into account in order to correct the SIF [50]

$$K_I = \sigma\sqrt{\pi a}\sqrt{\frac{2b}{\pi a}\tan\frac{\pi a}{2b}}\frac{0.752 + 2.02\frac{a}{b} + 0.37\left(1 - sin\frac{\pi a}{2b}\right)^3}{cos\frac{\pi a}{2b}} \tag{2.17}$$

Figure 2.9 shows the elastic crack tip stress distribution in front of the crack tip. The above equation is valid only in the crack tip region in which the stresses are dominated by the singularity.

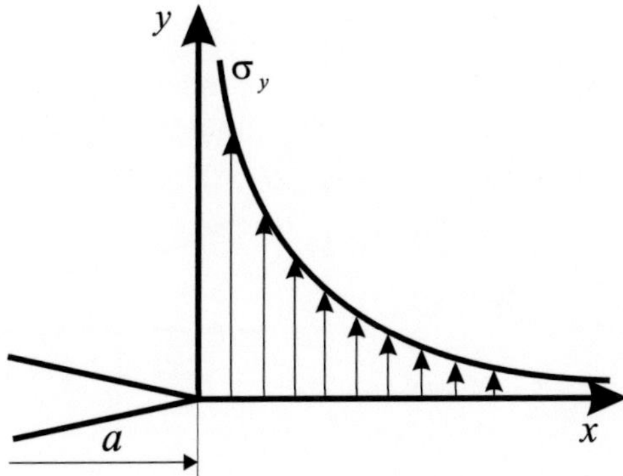

Figure 2.9: Elastic stress distributions at the crack tip

2.4.2.2 Elastic-plastic fracture mechanics (EPFM)

A. Irwin's model [61]

Irwin presented a simplified model for the determination of the plastic zone surrounding the crack tip under small-scale yielding. He focused only on the extent along the crack axis and not on the shape of the plastic zone. As a result of the crack tip plasticity, the displacements are larger and the stiffness of the plate is lower than that for the elastic case.

The length of the plastic zone c in front of the crack for plane stress is given by

$$c = \frac{1}{\pi} \left(\frac{K_I}{\sigma_Y} \right)^2 \tag{2.18}$$

where σ_Y is the yield stress. The effective crack has a length

$$a_{eff} = a + \frac{1}{2}c \tag{2.19}$$

Thus, the effective SIF for a effective length of crack a_{eff} is

$$K_I = \sigma\sqrt{\pi a_{eff}}\sqrt{\frac{2b}{\pi a_{eff}}tan\frac{\pi a_{eff}}{2b}}\frac{0.752 + 2.02\frac{a_{eff}}{b} + 0.37\left(1 - sin\frac{\pi a_{eff}}{2b}\right)^3}{cos\frac{\pi a_{eff}}{2b}} \tag{2.20}$$

An iterative process is needed for determining the effective SIF according to Irwin's model. Finally, the stress distribution directly ahead of the crack σ_y is calculated from the following equation

$$\sigma_y = \frac{\sigma\sqrt{\pi a_{eff}}}{\sqrt{2\pi x}} = \frac{K_{I\,eff}}{\sqrt{2\pi x}} \tag{2.21}$$

Figure 2.10 shows the elastic crack tip stress distribution in front of the crack tip based on Irwin's plastic zone correction.

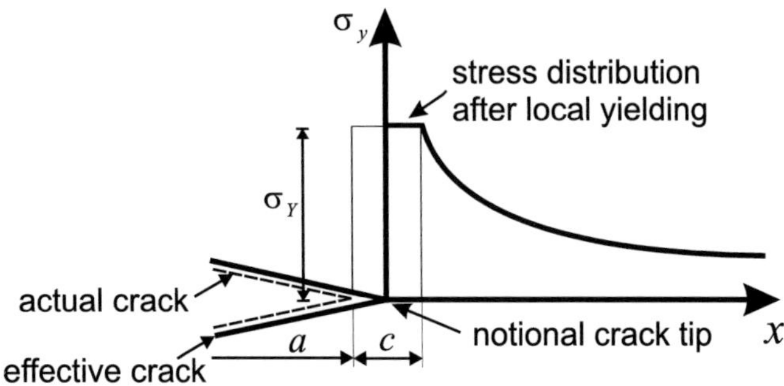

Figure 2.10: Stress distribution of Irwin's model ahead of the crack tip

B. Dugdale's model [37]

A simplified model for plane stress yielding which avoids the complexities of a true elasto-plastic solution was introduced by Dugdale [37]. The model applies to very thin plates in which plane stress conditions dominate, and to materials with elastic-perfectly plastic behaviour which obey the Tresca yield criterion.

Length of the plastic zone c is determined from the condition at which the stress distribution at the tip of the effective crack should remain bounded and equal to the yield stress. Length c in front of the physical crack carries the yield stress σ_Y tending to close the crack. (The part c is not really cracked, the material can still bear the yield stress [17]). Size of c is proposed such that the stress singularity disappears. Figure 2.11 presents the elastic crack tip stress distribution in front of the crack tip based on Dugdale's plastic zone model.

$$c = \frac{\pi}{8}\left(\frac{K_I}{\sigma_Y}\right)^2 \tag{2.22}$$

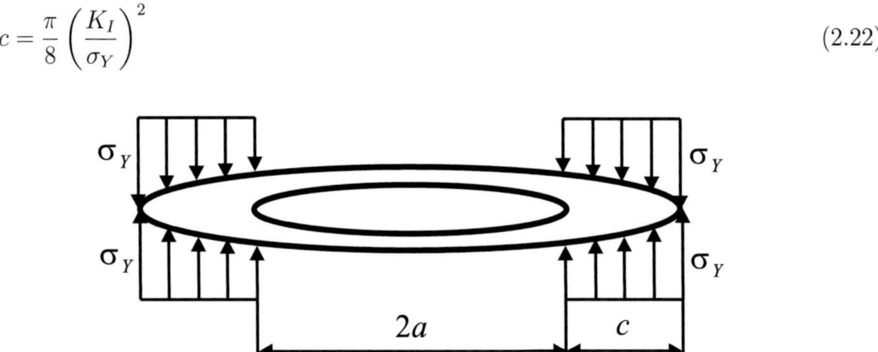

Figure 2.11: Dugdale's approach ahead of the crack tip

C. Cohesive zone model

In the FE approach, the CZM can be implemented as interface elements that are compatible with solid FEs. The cohesive elements are composed with zero thickness connected together with four-node solid elements (Figure 2.12).

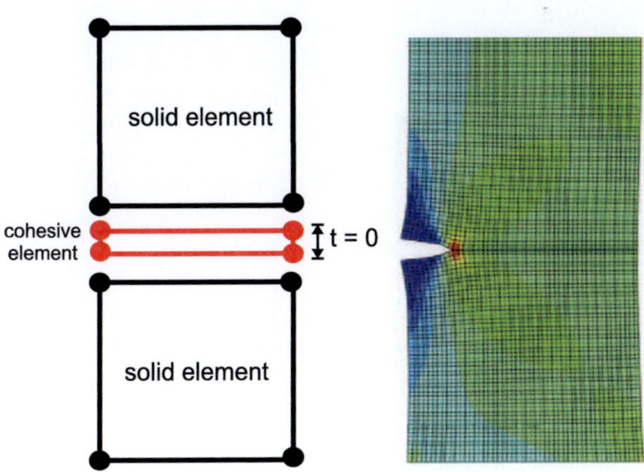

Figure 2.12: Schematic of an edge crack using CZM in FE model

For the sake of comparison, the elastic-plastic stress distributions based on Irwin's model and Dugdale's model are shown in Figure 2.13. As results, the Irwin's model underestimates the length of plastic zone by about 20%, compared to the Dugdale's model, but the CZM gives rather different results which depend on the plasticity parameter of material such as hardening exponential and shape of TSL. Figure 2.14 depicts the variation of the dimensionless quantities K_{eff}/K_I and σ_y/σ_Y using different approaches.

In the presence of CZM elements, the normal stresses drop in the region where normal separation has exceeded critical separation causing softening. When the separation between cohesive surfaces reaches maximum value for elements nearest to the crack tip, the stresses developed are lower within the softening zone. For distances beyond the softening zone, the presence of CZM elements does not influence the stress field as shown in Figure 2.13.

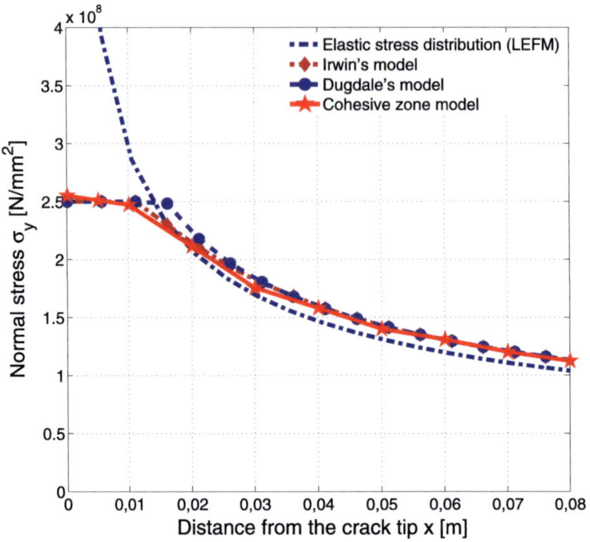

Figure 2.13: Stress distributions in front of the crack tip using LEFM, EPFM and CZM [82]

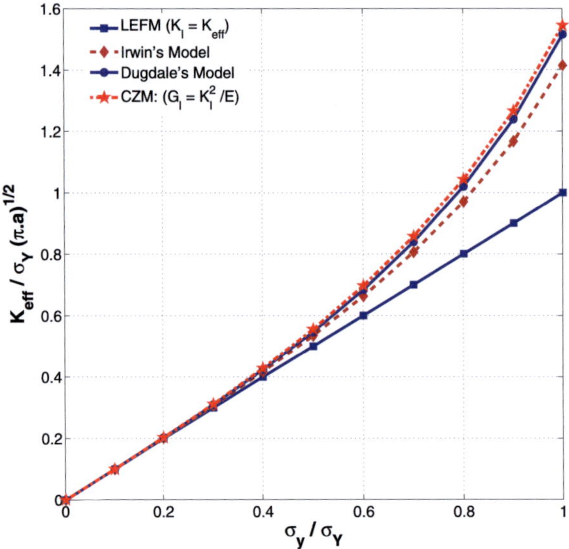

Figure 2.14: Normalised stress intensity factor versus normalised applied stress

2.4.3 Simulation of delamination

Delamination is a mode of failure for composite materials. Delamination, or interfacial cracking between composite layers, is one of the most common types of damage in laminated fibre-reinforced composites due to their relatively weak interlaminar strengths [144]. Eight-node cohesive elements are used to simulate mode-I crack growth on the double cantilever beam (DCB) and mode I-II crack growth on the mixed-mode bending (MMB). Both simulated models are 100 mm-long, 25 mm-wide and 2 x 1.5 mm-thick arms. The interface elements open according to some decohesion law and finally loose their stiffness so that the adjacent continuum elements are disconnected. Thus, the crack can propagate only along the element boundaries. If the direction of crack extension is not known in advance, the finite element mesh has to allow for different crack paths.

Cohesive FEs have been developed to capture the initiation and propagation of delamination cracks uses a bilinear relation between traction and separation or displacement jump. A bilinear TSL is similar to the softening law of the CZM but with an initial linear elastic response before crack initiation. This linear elastic part is defined using a penalty or interface stiffness parameter K_p, that ensures a stiff connection between the surfaces of the material discontinuity. The penalty stiffness should be large enough to provide connections but small enough to avoid numerical problem in a FE analysis. A reasonable choice for the value of penalty stiffness suggested by Zou *et al.* [160] is as follows

$$K_p = \left(10^5 \div 10^7\right) \tau_{max} \tag{2.23}$$

where τ_{max} is shear strength in N/mm^2 and the values of $(10^5 \div 10^7)$ in mm^{-1}.

Turon *et al.* [144] proposed the penalty stiffness of the CZM for delamination

$$K_p = \frac{\alpha_p E}{t} \tag{2.24}$$

where α_p is parameter much larger than 1 but smaller than 50, E is modulus of elasticity and t is thickness of the material.

If the interface is represented using a penalty stiffness parameter, the tractions can be reasonable obtained before damage initiates, although the separation or relative displacement do not have any physical meaning, as they can be affected significantly by the initial stiffness of the interface. Special attention must be paid to the choice of the traction threshold, i.e. the maximum value of the traction σ_{max} on the traction-separation curve for a single mode. It has been demonstrated in literature that, within a relatively large range, the maximum traction has little effect on the prediction of crack growth. However, it may influence the computational efficiency of a FE simulation.

In contrary, Shet and Chandra [130] reported that the maximum tractions significantly affect the crack growth, namely due to plasticity around the fracture process zone. Usually, the higher the traction threshold is, the more refined must be the mesh around the crack front and smaller load increments are required in order to avoid numerical problems. The load displacement curves obtained in the simulation are shown in Figure 2.15 together with a comparison with experimental results. It can be seen that the FE predictions almost match the experimental results.

Results presented in Figure 2.15 show that there are some differences between numerical and experimental results at the region of maximum displacement, which may be caused due to penalty stiffness, length of cohesive elements and also mesh size. The numerical solution of the DCB and MMB using CZM is directly related to the mesh characteristic. The coarser mesh size tends equilibrium points in harsh snap backs. Consequently, finer mesh size might results in the higher computational duration. Beside the mesh size, convergence is greatly being affected by the penalty stiffness. The penalty stiffness should be large enough to provide a reasonable stiffness but small enough to reduce the risk of numerical problems such as spurious oscillations of the tractions in an element.

2.5 Energy balance concept

When fracture proceeds, energy must be supplied by external loads. The bounding material undergoes elasto-plastic deformation involving elastic energy and plastic dissipative energy. In addition to plasticity, energy is supplied to the fracture process zone in form of cohesive energy that is dissipated within the cohesive elements. The cohesive energy is the sum of the surface energy and all dissipative processes that take place within the crack tip regime. For the present problem, a perfect energy balance between external work W and the sum of elastic energy E_{el}, plastic dissipative energy E_{pl}, cohesive energy E_{coh} and any other inelastic energy E_{inel}, for example damage/void growth will be assumed [130]. The energy balance is given by

$$W = E_{el} + E_{pl} + E_{coh} + E_{inel}$$

$$W = \int_0^t \left(\int_V \sigma : \dot{\varepsilon}_{el} \, \mathrm{d}V \right) \mathrm{d}t + \int_0^t \left(\int_V \sigma : \dot{\varepsilon}_{pl} \, \mathrm{d}V \right) \mathrm{d}t + \int_0^t \left(\int_S \sigma_n : \dot{\delta} \, \mathrm{d}S \right) \mathrm{d}t + E_{inel} \quad (2.25)$$

where σ, $\dot{\varepsilon}_{el}$, $\dot{\varepsilon}_{pl}$, σ_n and $\dot{\delta}$ are nominal stress tensor, elastic strain rate, plastic strain rate, cohesive traction and cohesive separation rate, respectively (including the terms for specimen volume V and the internal specimen surface S). The external work due to applied force is given by

$$W = \int_0^t \left(\int_S \mathbf{t} \cdot \mathbf{v} \, \mathrm{d}S \right) \mathrm{d}t + \int_0^t \left(\int_V \mathbf{b} \cdot \mathbf{v} \, \mathrm{d}V \right) \mathrm{d}t \quad (2.26)$$

where $\mathbf{v}$ is velocity field vector, $\mathbf{t}$ is exterior surface traction vector and $\mathbf{b}$ is body force vector.

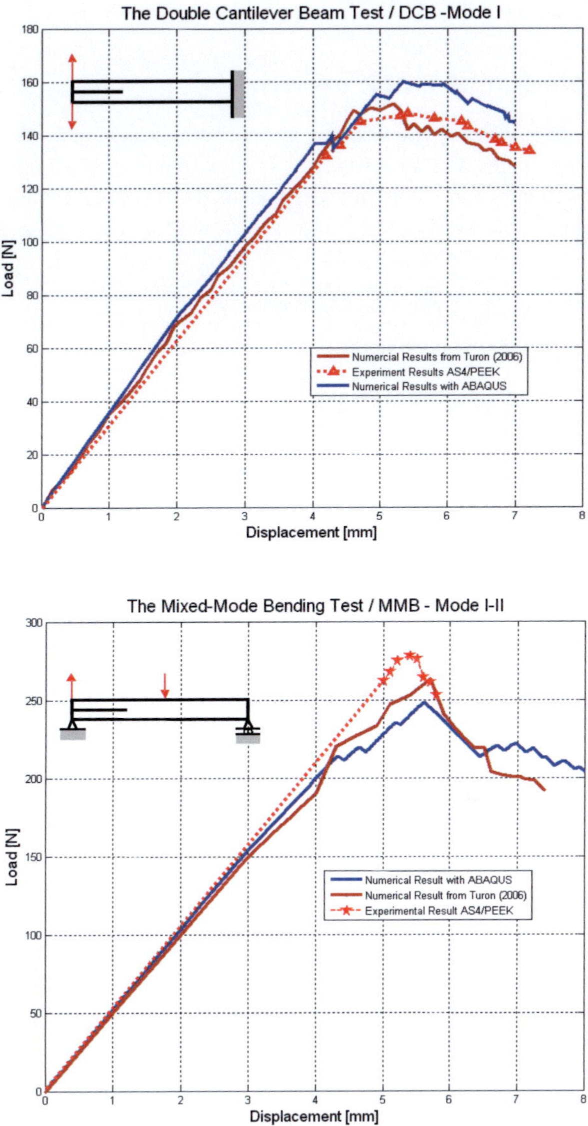

Figure 2.15: Numerical simulation and experimental data [142] for a Mode-I DCB test and mixed mode I-II MMB test

During FE analysis of crack growth, the amount of external work, elastic work, plastic work and other dissipative work are calculated. It is always seen that energy balance given by Eq.(2.25) is maintained in all FE computations.

2.5.1 Simulation of crack growth

As a first test case, a crack growth of a simple two-dimensional plate of width $b = 10$ cm and thickness $t = 1$ cm will be analyzed. The plate contains an edge crack of length $a_0 = 20$ mm and is subjected to a constant displacement in perpendicular direction to the crack plane. The plate is made from steel with physical properties given in Table 2.1.

Table 2.1: Mechanical properties of steel

σ_U	σ_Y	E	ν	ρ	K_{IC}
MPa	MPa	GPa	-	kg/m^3	MPa $\sqrt{m}$
400	250	210	0.30	7 850	66

The material of the plate is characterized by an elasto-plastic model. The uniaxial stress-strain response of the material is specified by

$$\sigma = \sigma_Y + K\varepsilon^n \tag{2.27}$$

where σ_Y is yield strength, K and n are the strength coefficient and strain hardening exponent, respectively.

The cohesive energy G_{IC} is usually taken to be the fracture toughness of the material, which in the present case is based on the value K_{IC} with a corresponding energy of

$$G_{IC} = \frac{K_{IC}^2}{E}\left(1 - \nu^2\right) \tag{2.28}$$

The penalty stiffness parameter K_p that ensures a stiff connection between the continuum elements and the cohesive elements is defined by Eq.(2.23). However, large values of the penalty stiffness may cause numerical problems. A bilinear TSL is used and has mechanical properties shown in Table 2.2.

Table 2.2: Mechanical properties of cohesive element

$\sigma_U = \sigma_{max}$	G_{IC}	K_p	δ_{sep}	δ_0
MPa	J/m^2	N/m^3	μm	μm
400	19 231	105×10^{13}	96	0.4

Some investigators changed boundary conditions of the crack tip node directly to obtain a free or fixed node. A common approach used to change the boundary conditions consists of connecting two springs to each boundary node. For free nodes, the spring stiffness is set equals to zero, and for the fixed ones an extremely large values of stiffness is assigned. To overcome the numerical difficulties resulted from large values of stiffness, Wu and Ellyin [153] use truss elements composed of a linear elastic material for simulating the crack extension. They connect truss elements to boundary nodes and a pair of contact surfaces on the crack line.

In the present work, four node elements based on CZM concepts have been developed and implemented as user defined elements within ABAQUS. In order to study the distribution of elastic, plastic, and cohesive energies in the fracture process zone, 100 continuum elements are used along the crack propagation path, namely 1 mm-side element and 50 cohesive elements in 1 continuum element or 0.02 mm-side cohesive element. A linear multi point constraint (MPC) is used to connect 50 cohesive elements with 1 continuum element. This scheme has several advantages, first, it is simple and does not need an extra subroutine. Second, it is efficient, namely it does not need too many continuum elements. Furthermore, several cohesive elements can be released simultaneously. This scheme yields accurate results in energy balance as to be discussed later on. The energy balance results obtained from the simulation are shown in Table 2.3.

Table 2.3: Energy balance for a single-edge crack plate

External work $W\,[J]$	Elastic strain energy $E_{el}\,[J]$	Plastic dissipative energy $E_{pl}\,[J]$	Cohesive energy $E_{coh}\,[J]$
13.959 40	13.579 00	0.079 93	0.300 48

This fracture energy is dissipative in nature, hence in an analysis using CZM, even for an elastic material the entire fracture energy of $J_{IC} = G_{IC}$ is dissipated through cohesive elements. For the case of elasto-plastic material, two distinct dissipation mechanisms can be identified, one due to plasticity within the bounding material, and another due to micro-separation processes in the fracture process zone. Note that

- Using Eq.(2.28), the cohesive energy is $G_{IC} = 19\,231$ J/m^2.

- Thus, the crack growth of one continuum element (1 mm-long and 1 cm thick) needs energy G_{IC} 1element $= 0.192\,31$ J.

- The length of crack for cohesive energy obtained from FE results (Table 2.3) is (0.300 48 J) / (0.192 31 J/mm^2)×(1 mm length of element) $= 1$ mm with rest of energy 0.108 17 J, which is used by the neighbour from cracked cohesive elements.

- Hence the length of crack is $a_0 + 1$ mm $= 21$ mm. The rest of the energy is used for 'partial crack growth' on a length of 5 mm by neighbour cohesive elements as shown in Figure 2.16.

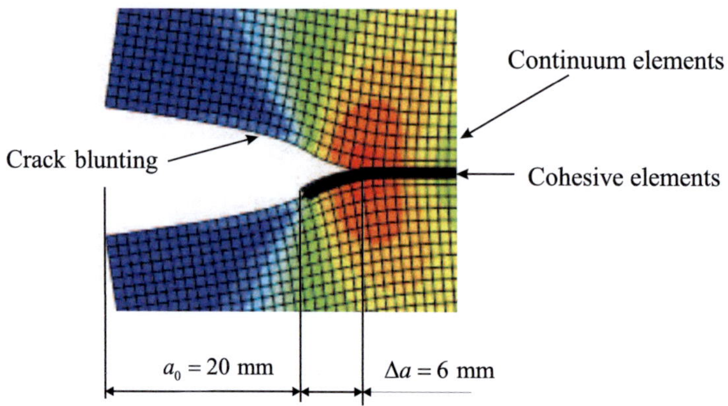

Figure 2.16: Geometry of the crack growth using cohesive elements

This calculation is very approximate, since yielding leads to stress redistribution and modifies the size and shape of the plastic zone. Initially the crack blunts due to deformation at the crack tip. Eventually, the material at the tip of the crack fails and the crack begins to advance. The initiation point is often difficult to determine. Furthermore, the approach of cohesive energy and critical SIF using Eq.(2.28) is commonly used for brittle material.

In order to be able to compare the results, let us consider the case of LEFM. The SIF K_I for a single edge crack plate is determined from Eq.(2.20).

- Using the cohesive energy 30 048 J/m^2 (0.300 48 J/(0.001 m-length of element $\times$ 0.01 m-thickness of element)), the SIF can be calculated as $K_I = 82.5 \times 10^6$ MPa$\sqrt{m}$.

- Using iterations with tolerance of crack length 0.000 1 mm, the effective length of crack a_{eff} computed from iterations Eq.(2.20) is 26.5 mm.

2.5.2 Energy distribution in purely elastic material

The numerical results indicates that during fracture growth process from initial crack length $a_0 = 20$ mm until $a = 28.4$ mm ($b = 10$ cm), recoverable elastic work E_{el} constitutes most of the external work, ranging from 96.5 to 99.5%, as shown in Figure 2.17.

2.5.3 Energy distribution in elasto-plastic material

Since elastic work is used in the entire body and not just for the crack tip region, this work depends on the geometry of the body. The plastic work occurs primarily in the crack tip region and hence is influenced by cohesive parameters. Though there are many parameters that can affect plasticity, the strain hardening exponent n and the cohesive strength σ_{max} (where $\sigma_{max} > \sigma_Y$) have the major influence. In general, the plastic work can be represented as

$$E_{pl} = f\left(\sigma_{max}, n\right) \tag{2.29}$$

Figure 2.18 represents the relationship between elastic strain energy, plastic dissipative energy and cohesive energy compared with the external work for $\sigma_{max}/\sigma_Y = 400/250 = 1.6$. The plastic energy represents about 0.7% of the overall energy.

Figure 2.19 depicts the relationship between plastic dissipative energy and cohesive energy compared with the external work for different values of σ_{max}/σ_Y namely, 1.6, 1.8, 2.0 and 2.5. For the value $\sigma_{max}/\sigma_Y = 1.6$ which represents small scale plasticity, the plastic energy represents about 25% of the overall energy dissipated. In other words, the error encountered when plastic work is not accounted for in the dissipative processes is of the order of 25% when small scale plasticity is observed. Increasing levels of plastic energy as a part of total dissipation when the values of $\sigma_{max}/\sigma_Y \geq 1.8$ gives significantly higher and is almost 100 to 250% as that of the cohesive energy. These results show that during crack growth with significant levels of plasticity, along amount of the external work will be used for plastic dissipative energy and the rest for the fracture process as cohesive energy.

The second parameter that affects the energy distribution during fracture process is the strain hardening exponent n. It should be noted that the hardening exponent is the parameter used to describe the hardening behaviour of the bounding material outside of the cohesive zone. In these simulations the crack is modelled using $\sigma_{max}/\sigma_Y = 1.6$. The value of n is selected as 0.26, 0.30, 0.40 and 0.50 to study the effect of n on the distribution plasticity dissipative energy and cohesive energy. Figure 2.20 gives information for all values of n, a small plastic zone is always generated before the crack initiates. Increased the strain hardening exponent tends to increase the plastic dissipative energy and reduces the cohesive energy.

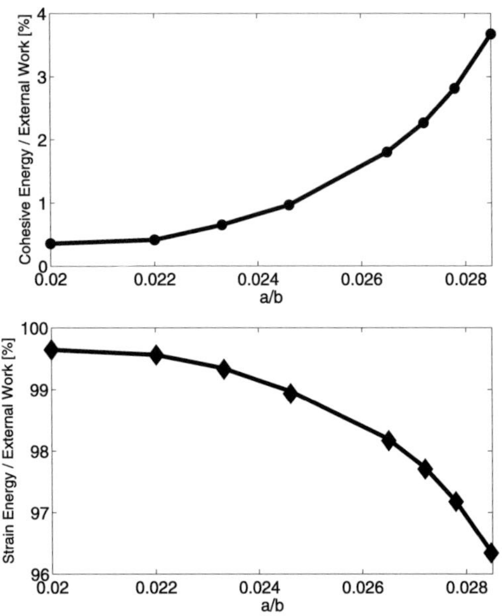

Figure 2.17: Variation of elastic strain energy and cohesive energy for pure elastic material

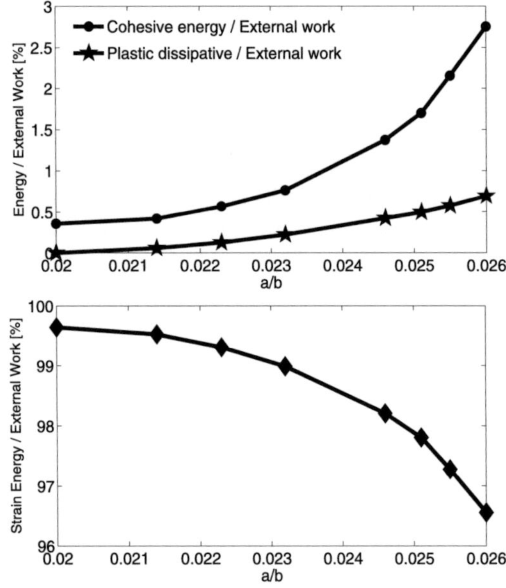

Figure 2.18: Variation of elastic strain energy, plastic dissipative energy and cohesive energy for elasto-plastic material

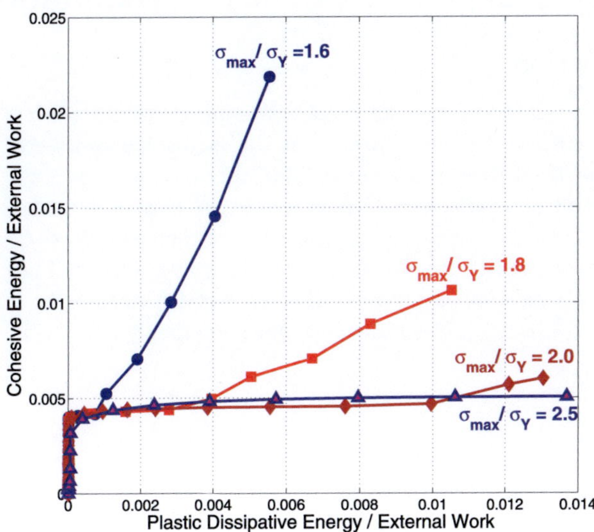

Figure 2.19: Variation of plastic energy and cohesive energy for different σ_{max}/σ_Y ratio

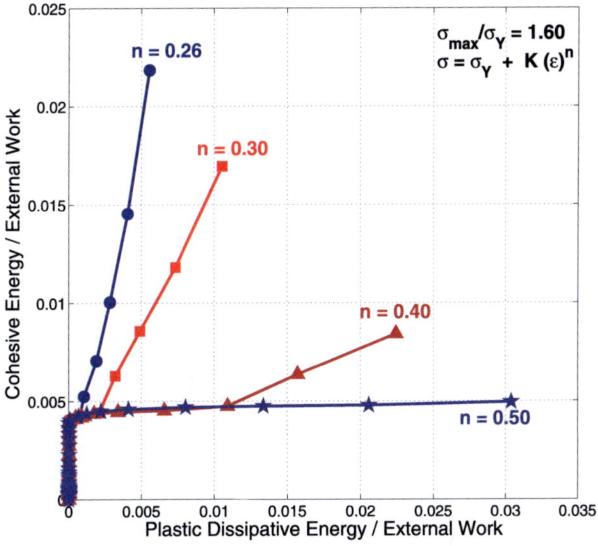

Figure 2.20: Variation of plastic energy and cohesive energy for different strain hardening exponent n

2.6 Triaxiality dependent cohesive zone model

2.6.1 Introduction

The main failure mechanism in ductile metals consists of the nucleation of voids and their growth and coalescence that initiates at the inclusions and second phase particles. Central to the growth of these voids is the triaxiality of the stress state known to greatly influence the amount of plastic strain which a material may undergo before ductile failure occurs. In ductile fracture of structures, triaxiality of the stress state depends on whether the structure is un-notched, notched or pre-cracked, as shown in Figure 2.21 [13]. The triaxiality of the stress state is given as the ratio of hydrostatic or mean stress to the effective stress (von Mises equivalent stress), mathematically

$$\chi = \frac{\sigma_m}{\sigma_{eq}} \tag{2.30}$$

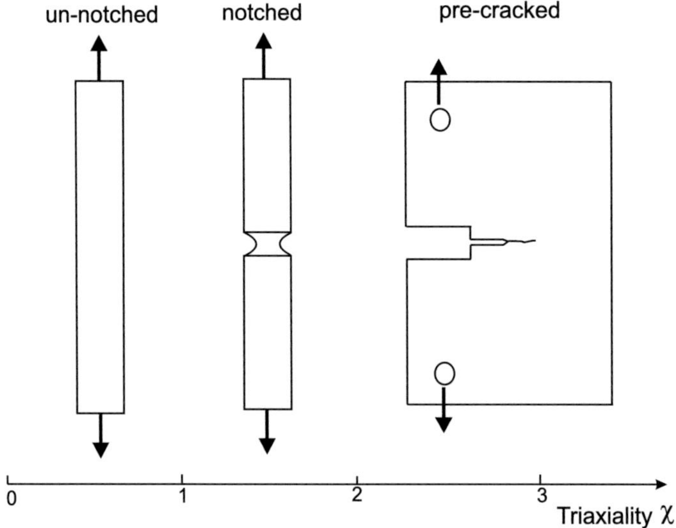

Figure 2.21: Triaxiality parameter χ at initiation of ductile fracture in different geometries

If material failure is due to ductile fracture, i.e. induced by void growth and void coalescence, there exist two approaches to model the material separation:

- One approach that can reflect the local of stress carrying capacity is the introduction of a CZM where the tractions in the process zone are related to the magnitude of material separation. Hillerborg et al. [56] and Needleman [94] incorporated the cohesive zone concept in the FEM by devising CZM's which obeyed an assumed TSL.

- Ductile crack growth due to nucleation of voids and subsequent growth and coalescence has been analyzed by a porous plasticity model (GTN model) introduced by Gurson and improved by Tvergaard and Hutchinson [145]. The porous material model uses an approximate yield condition which depends on the triaxiality of the stress state as well as the current void volume fraction. Material failure is assumed to occur at a critical value of void volume fraction. The processes of nucleation, growth and coalescence of voids, each require a set of parameters which may not be unique. A commonly encountered problem models based on micro-mechanics is that, even though they provide into insight the basic physics of the problem they involve far too many parameters to describe the macroscopic behaviour, and hence, are difficult to apply for practical purposes.

Siegmund and Brocks [133] investigate debonding of two elastic-plastic blocks by a ductile failure mechanism. The bond line between the two blocks is specified either as a strip of a material described by the modified Gurson constitutive equation or by a CZM. Special consideration in their paper is given to the influence of the triaxiality of the stress state on the cohesive material parameters. The process zone bonding the two blocks together is described in two different ways as shown in Figure 2.22.

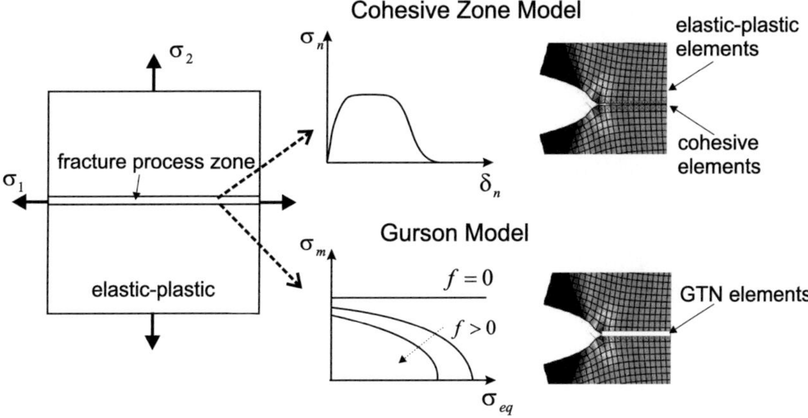

Figure 2.22: Visualization of the two approaches used for description of the bonding region

2.6.2 Constitutive behaviour of the continuum (undamaged material)

The mechanical behaviour of the bulk material is independent of the softening function and can be described by an unrestricted constitutive equation. The parent material or continuum is taken to be power law strain hardening such that the uniaxial stress-strain curve is represented by

$$\varepsilon = \begin{cases} \frac{\sigma}{E} & \text{for } \sigma \leq \sigma_Y, \\ \frac{\sigma_Y}{E}\left(\frac{\sigma}{\sigma_Y}\right)^{\frac{1}{n}} & \text{for } \sigma > \sigma_Y. \end{cases} \tag{2.31}$$

where E is the modulus of elasticity, σ_Y the initial yield stress and n is the strain hardening exponent.

Ideally, the model within the cohesive zone should be able to replicate the constitutive behaviour of the undamaged material: linear elastic followed by strain hardening, until the conditions for a softening process due to damage are reached. The softening process representative of increasing material degradation is triggered by rapid growth of voids as consequence of highly triaxial state of stress. The model is proposed for ductile metals in which the fracture process is localized in a thin layer which has thickness of a void spacing (Siegmund and Brocks [134], Anvari *et al.* [6], Scheider [118], Banerjee und Manivasagam [13]).

The TSL is linear up to the separation limit δ_1, exhibits strain hardening up to δ_2 and softening as shown in Figure 2.23. Each part of the curve as well as the limiting separation parameter δ_2, are dependent on the triaxiality parameter χ as expressed in the next section.

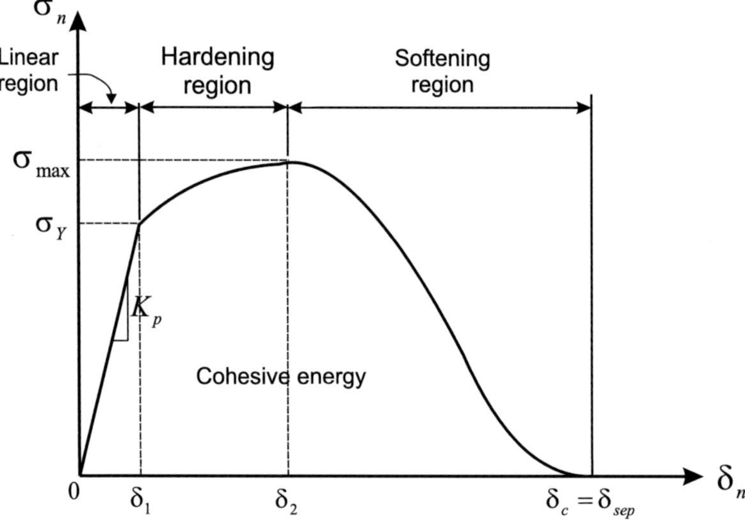

Figure 2.23: Traction separation law for ductile materials

2.6.3 Formulation of triaxiality dependent Model

This section is concerned with crack growth in elasto-plastic materials that exhibit ductile fracture, i.e. induced by void growth and void coalescence. There exist two approaches to model material failure [118]

- Gurson-Tvergaard-Needleman (GTN) model [51], [145]), [95]

- Cohesive zone model (CZM) [37], [14] with
    - constant parameters (numerically fitted to the experiments) or
    - triaxiality dependent parameters (determined from GTN unit cell)

The triaxiality of stress state as ratio of hydrostatic stress (mean normal stress) to the effective stress (von Mises equivalent stress) from Eq.(2.30), under conditions of plane strain and developed during elastic deformation can be calculated from a elastic constitutive relationship to be

$$\chi_{el} = \frac{\frac{1}{3}\left(\sigma_1 + \sigma_2 + \sigma_3\right)}{\frac{1}{\sqrt{2}}\sqrt{\left(\sigma_1 - \sigma_2\right)^2 + \left(\sigma_2 - \sigma_3\right)^2 + \left(\sigma_3 - \sigma_1\right)^2}} \tag{2.32}$$

where σ_1, σ_2 and σ_3 are the first, second and third principal stresses, respectively.

For a state of plane strain,

$$\sigma_3 = \nu\left(\sigma_1 + \sigma_2\right) \tag{2.33}$$

and the stress ratio,

$$r_\sigma = \frac{\sigma_1}{\sigma_2} \tag{2.34}$$

Substituting Eqs.(2.33) and (2.34) into Eq.(2.32) yields

$$\chi_{el} = \frac{(1+\nu)(1+r_\sigma)}{3\sqrt{(r_\sigma^2 + 1)(\nu^2 - \nu + 1) + r_\sigma\left(2\nu^2 - 2\nu - 1\right)}} \tag{2.35}$$

where ν is Poisson's ratio. In the material limit of incompressible deformation ($\nu = 0.5$), the triaxiality parameter reaches its extreme value from Eq.(2.35)

$$\chi_{pl} = \frac{1 + r_\sigma}{\sqrt{3}\left(1 - r_\sigma\right)} \tag{2.36}$$

Hancock and Brown [54], [53] have investigated fracture of different geometries, for which there is a single failure locus of the effective plastic strain $\bar{\varepsilon}^{pl}$ at fracture initiation as a function of the triaxiality parameter. It was found that represents ductile fracture of all different specimens.

The triaxiality dependent failure locus is given by

$$\bar{\varepsilon}^{pl} = K e^{-\frac{3}{2}\chi} \tag{2.37}$$

where K is a material dependent non-dimensional parameter.

For any deformation beyond initial yield the triaxiality in absence of any damage to the material would increase from the elastic value to its extreme value χ_{pl}. However, at large ratios ($r_\sigma > 0.5$), after the onset of yield as the triaxiality χ increases from its elastic limit, the void nucleation and growth would correspondingly increase. This would imply that the extreme value of triaxiality χ_{pl}, is never realized as the material will fail at a lower value of triaxility χ. To incorporate this effect, Banerjee and Manivasagam [13] propose a saturation limit of the triaxiality parameter χ_{sat}, for which the equivalent plastic strain for failure due to void growth is

$$\chi_{sat} = -\frac{2}{3} \ln \left(\frac{\bar{\varepsilon}^{pl}}{K} \right) \tag{2.38}$$

$$= -\frac{2}{3} \ln \left(\frac{S\sigma_Y}{CE} \right) \tag{2.39}$$

where E is the modulus of elasticity, σ_Y the initial yield stress, S a non dimensional multiplicative factor of the elastic strain at yield and C is a non dimensional material constant of the order of the material constant K.

Banerjee and Manivasagam [13] propose that the ductile failure of material is not at the extreme value of the triaxiality parameter χ_{pl}, but at an effective triaxiality parameter χ_{eff}. The effective triaxiality parameter is taken to be such that at low stress ratio. It follows χ_{pl} while for higher stress ratio it saturates to the saturation limit of Eq.(2.39), and is of the form

$$\chi_{eff} = \frac{1}{2}(\chi_{pl} + \chi_{sat}) - \frac{1}{2}(\chi_{pl} - \chi_{sat}) \tanh \left(\frac{r_\sigma - r_{\sigma cr}}{0.1} \right) \tag{2.40}$$

where $r_{\sigma cr}$ is the critical value of stress ratio r_σ, which indicates the onset of the saturation of effective triaxiality calculated using Eq.(2.36), hence

$$\chi_{sat} = \frac{1 + r_{\sigma cr}}{\sqrt{3}\,(1 - r_{\sigma cr})}$$

$$\Leftrightarrow r_{\sigma cr} = \frac{\sqrt{3}\chi_{sat} - 1}{\sqrt{3}\chi_{sat} + 1} \tag{2.41}$$

For proportional loading in plane strain condition, the dependence of the triaxiality parameters χ_{el} and χ_{pl} on the stress ratio r_σ, is shown in Figure 2.24.

- For lower stress ratio ($r_\sigma < 0.3$), the difference between the triaxiality parameters in the elastic limit and plastic limit is not as significant as when r_σ tends to unity.

- While χ_{pl} tends to infinity, χ_{el} remains finite.

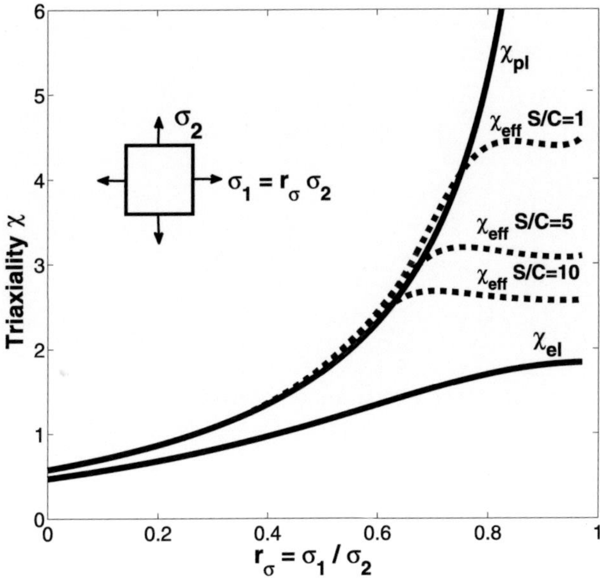

Figure 2.24: Dependence of triaxiality parameter on a fixed applied stress ratio under plain strain

- The effective triaxiality parameter follows the plastic limit of triaxiality until the stress ratio reaches the critical value $r_{\sigma cr}$, when it starts to saturate to χ_{sat}.

- Figure 2.24 also shows effective triaxiality at failure χ_{eff} depend on the ratio S/C.

2.6.4 Traction-Separation Law

Since the CZM is a phenomenological model, various formulations for defining the shape of the TSL and the cohesive parameters are in use [94], [155], [38], [130], [59]. Among these definitions, the one introduced by Scheider [118], Scheider *et al.* [119], [121], [120], [25] have been used to calculate damage and failure of the unit cell by a cohesive element. Their TSL has two shape parameters δ_1 and δ_2. In addition to the cohesive parameters already introduced consists of three parts: increasing, constant and decreasing traction (decohesion). The traction as function of the separation in their model is

$$\sigma_n = \sigma_{max} \begin{cases} 2\left(\frac{\delta}{\delta_1}\right) - \left(\frac{\delta}{\delta_1}\right)^2 & \text{for } 0 < \delta < \delta_1, \\ 1 & \text{for } \delta_1 < \delta < \delta_2, \\ 2\left(\frac{\delta-\delta_2}{\delta_c-\delta_2}\right)^3 - 3\left(\frac{\delta-\delta_2}{\delta_c-\delta_2}\right)^2 + 1 & \text{for } \delta_2 < \delta < \delta_c. \end{cases} \tag{2.42}$$

By changing δ_1 and δ_2, one can have a variety of TSL shapes (see Figure 2.25). The cohesive energy is written as

$$
G_I = \sigma_{max} \left[\int_0^{\delta_1} \left[2\left(\frac{\delta}{\delta_c}\right) - \left(\frac{\delta}{\delta_c}\right)^2 \right] \mathrm{d}\delta + \int_{\delta_1}^{\delta_2} \mathrm{d}\delta \right]
$$

$$
+ \sigma_{max} \left[\int_{\delta_2}^{\delta_c} \left[2\left(\frac{\delta - \delta_2}{\delta_c - \delta_2}\right)^3 - 3\left(\frac{\delta - \delta_2}{\delta_c - \delta_2}\right)^2 + 1 \right] \mathrm{d}\delta \right] \tag{2.43}
$$

$$
= \sigma_{max} \left(\frac{1}{2}\delta_c - \frac{1}{3}\delta_1 + \frac{1}{2}\delta_2 \right) \tag{2.44}
$$

In this formulation, the maximum separation occurs at

$$
\delta_c = \frac{2G_I}{\sigma_{max}} \frac{1}{1 - \frac{2\delta_1}{3\delta_c} + \frac{\delta_2}{\delta_c}} \tag{2.45}
$$

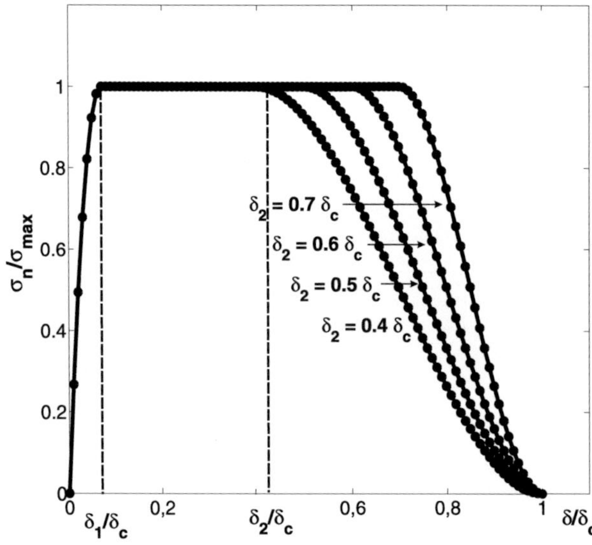

Figure 2.25: TSL shape proposed by Scheider [118]

Another formulation was introduced by Banerjee and Manivasagam [13]. They proposed a versatile cohesive zone model to predict ductile fracture at different states of stresses. The formulation developed for mode-I plane strain accounts for triaxiality of the stress-state explicitly by using basic elasto-plastic constitutive relations combined with two stress-state independent new model parameters. They compared their model with available predictions of CZMs based on porous plasticity damage models too, and proposed TSL that has three distinct regions of constitutive behaviour: the traction separation law is linear up to the separation limit δ_1, exhibits strain hardening up to δ_2 followed by softening curve:

$$\sigma_n = \begin{cases} \left(1 + \sqrt{3}\chi_{eff}\right)\frac{2E}{3}\delta & \text{for} \quad 0 < \delta < \delta_1, \\ \left(1 + \sqrt{3}\chi_{eff}\right)\frac{\sigma_Y}{\sqrt{3}}\left(\frac{2E}{\sqrt{3}\sigma_Y}\delta\right)^n & \text{for} \quad \delta_1 < \delta < \delta_2, \\ \sigma_{max}\exp\left(-0.01\left(\frac{\delta-\delta_2}{\delta_2}\right)^4\right) & \text{for} \quad \delta_2 < \delta < \delta_c. \end{cases} \tag{2.46}$$

The linear behaviour of the CZM exists until the separation limit defined by von Mises yield condition is reached

$$\delta_1 = \frac{\sqrt{3}\sigma_Y}{2E} \tag{2.47}$$

Further separation results in strain-hardening up to

$$\delta_2 = \frac{\sqrt{3}}{2}\left(Ce^{-\frac{3}{2}\chi_{eff}} + \frac{\sigma_Y}{E}\right) \tag{2.48}$$

As comparison with available predictions of CZMs based on porous plasticity damage models from Scheider [118], is shown in Figure 2.26.

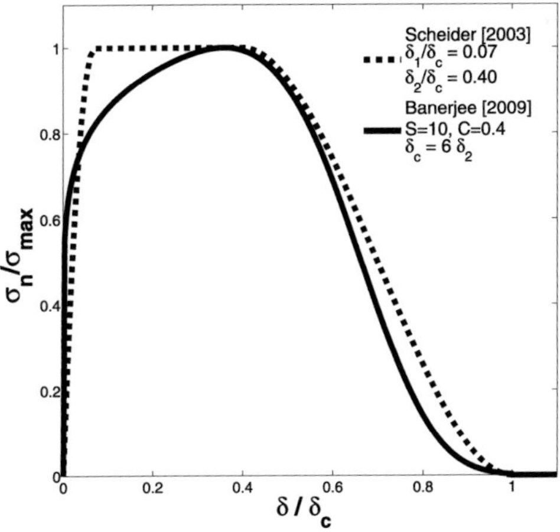

Figure 2.26: TSL shape proposed by Scheider [118] compared to Banerjee [13]

The difference of both proposed TSL shapes may be caused by their difference approaches. Scheider [118] used a simple 2D axisymmetric single-element test for the identification of the TSL and defined the specific GTN parameters while Banerjee and Manivasagam [13] proposed their TSL shapes explicitly by using basic elasto-plastic constitutive relations combined with two stress-state independent new model paramaters. One has to keep in mind that the validity of the GTN model is limited, first of all with respect to the failure mechanism, but also with respect to stress triaxiality. The known problems with low triaxialities, for example, make the proposed identification procedure only applicable for high constraint structures unless a more sophisticated void growth model is utilised. In the case that many layers of cohesive elements are introduced to the FE model, e.g. for the numerical prediction of the crack path, then the initial stiffness has a strong influence on the structural response.

The TSL proposed by Banerjee and Manivasagam [13] is shown in Figure 2.27 for steel and aluminium with the effect of the applied stress ratio in the range $r_\sigma = 0$, 0.1, 0.2, 0.4 and 0.6. In the figures, as the stress ratio is increased, the corresponding higher triaxiality of the stress state inhibits plastic deformation. As a result, onset of yielding is at a higher normal traction and the peak stress is higher as well. While higher triaxiality inhibits plastic deformation, it promotes void nucleation, growth and coalescence. Thus, the softening behaviour begins at lesser separation for higher stress ratios. In summary, for higher stress ratios the TSL curves tend to be steeper and narrower.

The effect of a non dimensional material constant (model parameter) C is shown in Figure 2.28 for steel. As the stress ratio of the applied stress-state increases the peak stress value rises up to a saturation value corresponding to r_σ. The effect of C is to increase the peak stress level for the entire range of stress ratio. The dependence of the cohesive parameters σ_{max}/σ_Y on stress triaxiality as depicted in Figure 2.29 is obtained from Eq.(2.46) and for cohesive energy is calculated from the area under the curve in Figure 2.27.

In summary, the formulation and validation of a versatile CZM proposed by Banerjee and Manivasagam [13] is well presented which incorporates triaxiality explicitly, while the conventional cohesive parameters, peak stress and cohesive energy are not material constants as they depend on the triaxiality of the stress state.

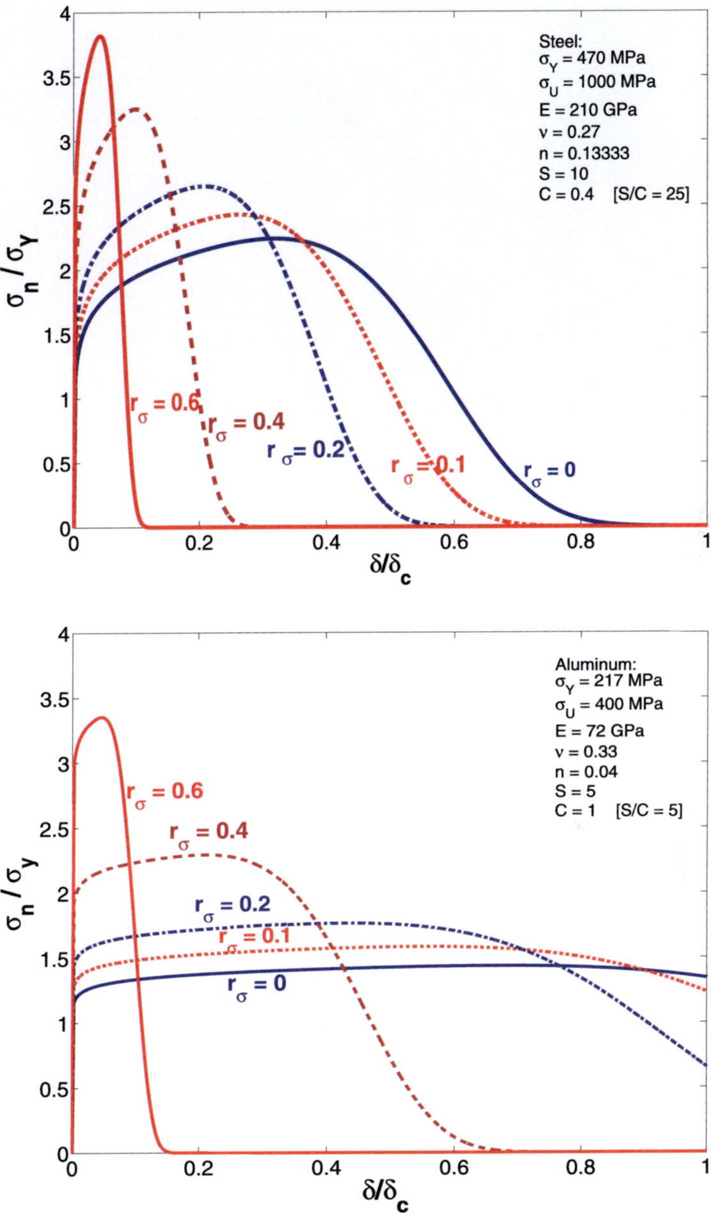

Figure 2.27: Traction-separation for different stress ratios r_σ for steel and aluminium

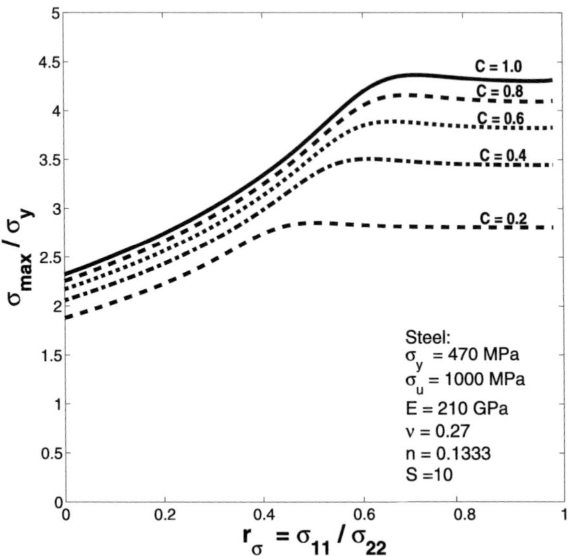

Figure 2.28: Effect of stress ratios r_σ on peak stress for steel

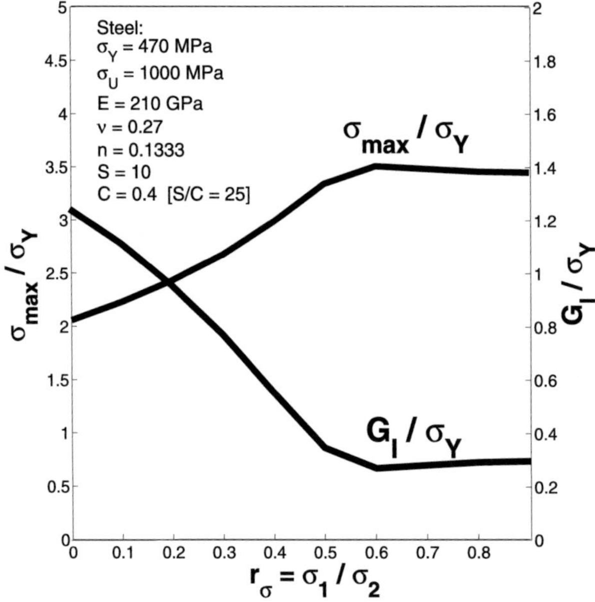

Figure 2.29: Dependence of the normalised cohesive parameters: cohesive strength σ_{max}/σ_Y and cohesive energy G_I/σ_Y on the triaxiality for steel

2.7 Implementation of the cohesive model for a cracked shaft

2.7.1 Classification of cracks

Based on geometry of cracks on a rotor, cracks can be broadly classified as follows [116]:

Transverse crack Cracks perpendicular to the shaft axis are known as transverse cracks. Most past and current research focuses on the detection of such cracks. They introduce a local flexibility in the stiffness of the shaft due to strain energy concentration in the vicinity of the crack tip.

Longitudinal crack Cracks parallel to the shaft axis are known as longitudinal cracks.

Slant crack Cracks at an angle to the shaft axis are known as slant or helicoidal cracks. These cracks are also encountered, but not very frequently. Slant cracks influence the torsional behaviour of the rotor in a manner quite similar to the effect of transverse cracks on the lateral behaviour. Their effect on lateral vibrations is less than that of transverse cracks of comparable severity.

Breathing crack Cracks that open when the affected part of the material is subjected to tensile stresses and close when the stress is reversed are known as breathing cracks. The stiffness of the component is most influenced when under tension. Shaft cracks breath when crack sizes are small, running speeds are low and radial forces are large.

Open or gaping crack Cracks that always remain open are known as gaping cracks. They are more correctly called notches.

Surface crack Cracks that open on the surface are called surfaces cracks. They can normally be detected by techniques such as dye-penetrant, or visual inspection.

2.7.2 Breathing crack under rotating load

In this section, the breathing crack of a circular cross-section shaft is presented using CZM. Figure 2.30 shows the isometric view of the model with a relative crack depth $a/d = 0.1$. Length and diameter of shaft are 1.0 m and 0.08 m, respectively. The breathing mechanism is generated by a bending load due to external load, and shown over one cycle 2π rad (360°) by increasing the angle by steps of $\pi/6$ rad (30°). The observation of opening crack is repeated for all different angular positions of the cracked shaft specimen. The breathing mechanism is observed by the nodes displacement around the crack.

The FE model of shaft shown in Figure 2.30 has more than 5 000 elements, have been used for the analysis of the cracked cylindrical beam. The cohesive elements along the crack surfaces are implemented as interface elements that are compatible with solid FEs. The CZM approach to model the breathing crack can be used accurately due to the fact that the crack tip is supposed to be formed by the boundary between the cracked areas and the

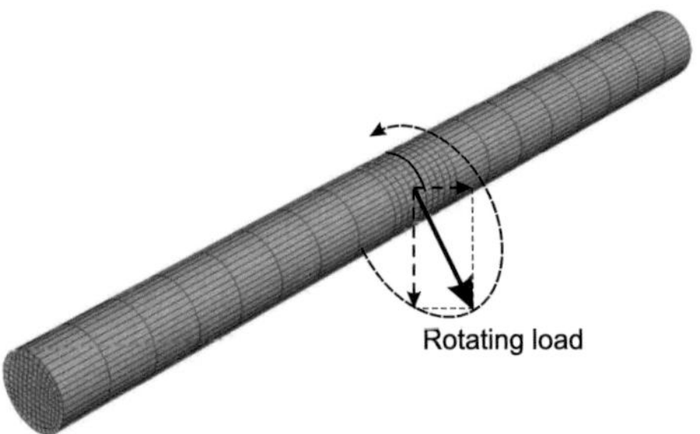

Figure 2.30: Breathing crack model of the non rotating cracked shaft subjected to rotating load

uncracked areas. It is correct when the crack is partially open by the boundary between the closed cracked and the open cracked areas because the CZM approach is based on the TSL. The cohesive energy will appear when the crack opens and will not appear when the crack closes.

Figure 2.31 represents breathing mechanism results obtained over one revolution of the rotating load. As can be seen the crack opens more slowly at the beginning, but increases its opening at $\pi/3$ rad (60°). It is more open and at $5\pi/6$ rad (150°) it is already completely open. The crack closes again at $4\pi/3$ rad (240°) and increases its closing at $3\pi/2$ rad (270°). The crack is already completely close at $11\pi/12$ rad (330°). Breathing crack results under rotating load (Figure 2.31) has limitation due to quasi static condition. These issues will be discussed detail in Chapter 6.

2.7.3 Breathing versus open crack: change in second moment of area

In order to analyse real cracked shaft behaviour, the open cracked models will be used to compare to the lower natural frequency of the breathing crack results approximated by FE model. This section represents the open crack model results approximated by using Mayes' model which represents the fractional change in the second moment of area as measured at the crack face. Mayes and Davis [90] related the change in second moment of area $(\Delta I/I)$ to the relative crack depth (a/d) as

$$\frac{\frac{\Delta I}{I}}{1 - \frac{\Delta I}{I}} = \frac{r}{L}\left(1 - \nu^2\right) F\left(\frac{a}{d}\right) \tag{2.49}$$

where r, L, I and ν are shaft radius, length of the section with reduced properties, second moment of area and Poisson's ratio, respectively. $F(a/d)$ is a function independent of all other parameters and is a universal function for a given shape of crack. In principle, the function $F(a/d)$ can be derived from the appropriate stress concentration factor.

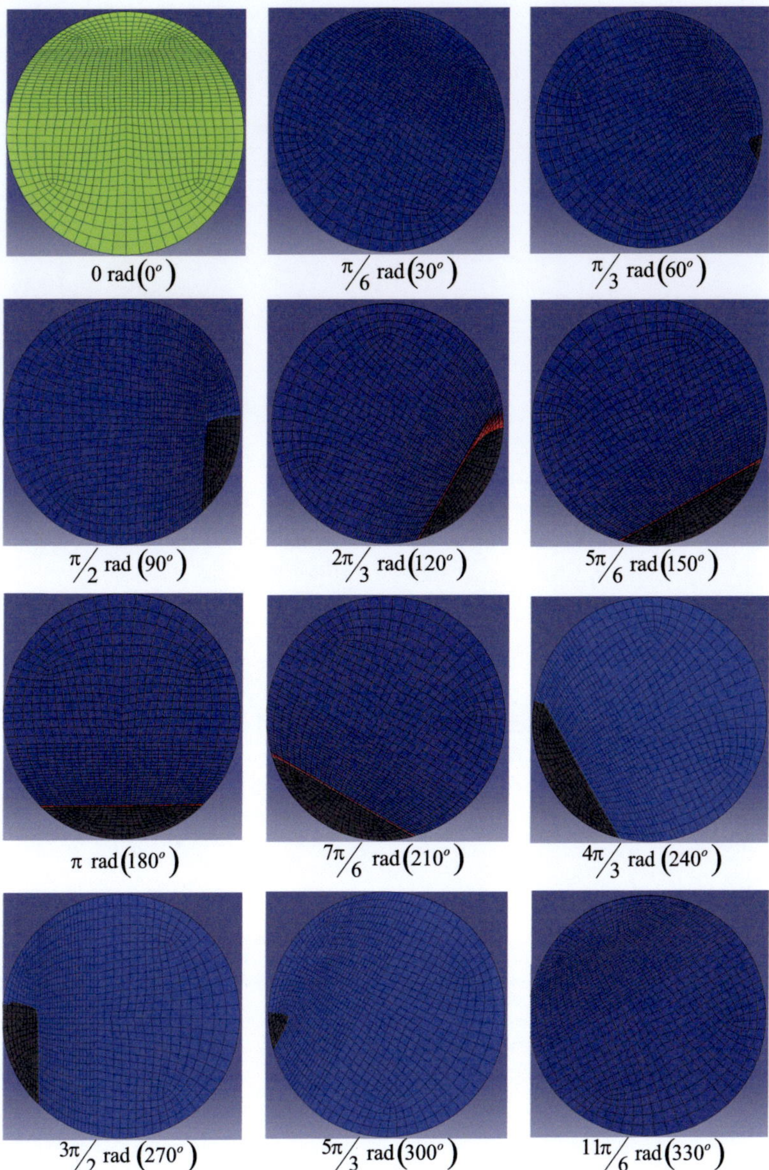

Figure 2.31: Breathing crack of the non rotating cracked shaft under rotating load for relative crack depth a/d=0.1

It was shown that to a very good approximation to this function is just equal to the fractional change in the second moment of area as measured at the crack face. $F(a/d)$ is a non linear function of relative crack depth a/d. A crack may be represented by reducing the second moment of area at the crack section by ΔI. From Eq.(2.49), the function $F(a/d)$ can be plotted as shown Figure 2.32 [77].

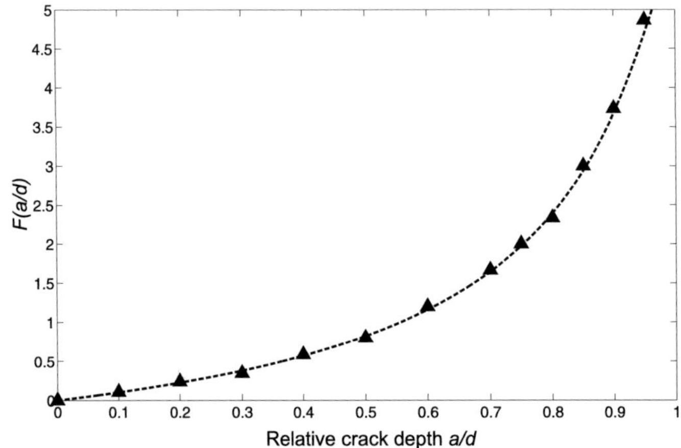

Figure 2.32: Variation of function $F(a/d)$ vs. relative crack depth a/d

Consider a simply supported cracked shaft with length 1.0 m and diameter 0.08 m. Material of shaft with Young's modulus, density and Poisson's ratio are 210 GPa, 7 850 kg/m^3 and 0.3, respectively. The relative crack depth $a/d = 0.1$ is in the middle of the shaft as shown in Figure 2.33.

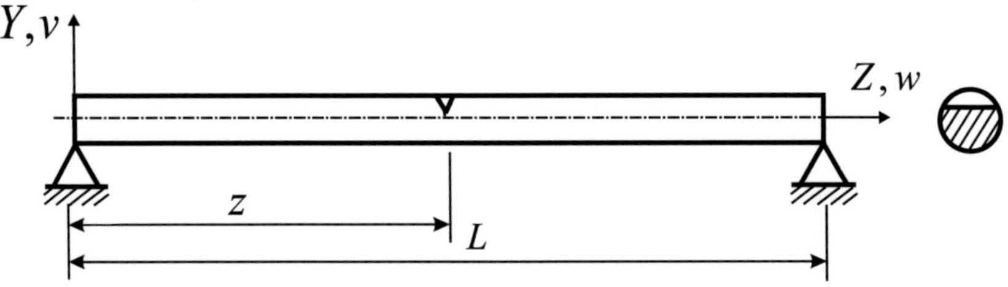

Figure 2.33: Rotor with an open crack simply supported on both ends

Using Figure 2.33 and Eq.(2.49), we obtain moment inertia of the open cracked shaft

$$I_{cr} = \frac{I}{1 + \frac{d}{2L}\left(1 - \nu^2\right) F\left(\frac{a}{d}\right)} \tag{2.50}$$

Using Euler-Bernoulli beam theory, the first three natural frequencies of

$$\omega_{cr_n} = \left(\frac{n\pi}{L}\right)^2 \sqrt{\frac{EI_{cr}}{\rho A}} \quad n = 1, 2, 3, \ldots \tag{2.51}$$

where for $a/d = 0.1$, $\Delta I/I = 0.003\,75$ or $I_{cr} = 2{,}0031 \times 10^{-6}$ m^4.

In order to compare the results using the change in second moment of area from Mayes' model, FE model using software ABAQUS is applied with the same material and geometry. Table 2.4 represents results of the natural frequency of uncracked and cracked shaft at rest.

Table 2.4: Natural frequencies of uncracked and cracked shaft a/d=0.1 at rest in [rad/s]

Mode	Uncracked shaft Theoretic	Uncracked shaft FEM	Cracked shaft Mayes' model	Cracked shaft FEM
1	1 020.9	1 019.7	1 019.1	1 017.8
2	4 083.8	4 081.3	4 076.2	4 068.5
3	9 188.5	9 123.3	9 171.4	9 164.2

2.7.4 Breathing versus open crack: change in local stiffness

The second approach is introduced by Dimarogonas and Papadopoulos [35]. They used local stiffness c_L that is approximated from dimensionless local flexibility $\bar{c}$ which varies with the relative crack depth. A continuous cracked shaft model is assumed with an open crack model. It was noted that, in case of a horizontal rotor, the dead weight bending plays an important role in the dynamics. At any stage of the cycle, the crack may be open or close depending on the operating conditions, indeed if operating very close to a critical speed, the crack may remain open throughout a full shaft revolution if the unbalance is very high. However, for modest levels of unbalance and away from critical speeds, the equations governing the shaft motion may be linearised by observing that the bending moment due to dead weight bending will dominate over inertial terms. It is assumed that only the open portions of the crack change the stiffness of rotors.

A continuous model is presented for vibration analysis with an open crack, which is based on assumptions that the cracked rotor is an Euler-Bernoulli beam with circular cross section and that the crack region is modelled as a local flexibility. The dimensionless local flexibility $\bar{c}$ is calculated with fracture mechanics method proposed by Dimarogonas and Papadopoulos [35]. Moreover, the explicit natural frequencies of an open cracked shaft based on fracture mechanics can be calculated by Euler-Bernoulli beam theory (Dong *et al.* [36], Dharmaraju *et al.* [32], Chondros *et al.* [24], [23] and Rizos *et al.* [114]. The continuous model of the cracked shaft is shown in Figure 2.33.

The equation of motion of a uniform shaft based on Euler-Bernoulli beam theory in the well known form

$$EI\frac{\partial^4 y(z,t)}{\partial z^4} + \rho A\frac{\partial^2 y(z,t)}{\partial t^2} = 0 \tag{2.52}$$

Using the separation of variables, the partial differential equation can be solved by introducing

$$y(z,t) = Y(z)\sin(\omega t) \tag{2.53}$$

Substituting, one obtains

$$Y^{iv}(\beta) - \lambda^4 Y(\beta) = 0 \tag{2.54}$$

where

$$\beta = \frac{z}{L} \tag{2.55}$$

$$\lambda^4 = \omega^2\frac{\rho AL^4}{EI} \tag{2.56}$$

The general solution is given by

$$Y_1(\beta) = A_1\cosh(\lambda\beta) + A_2\sinh(\lambda\beta) + A_3\cos(\lambda\beta) + A_4\sin(\lambda\beta) \quad \text{if } \beta \in [0,\beta) \tag{2.57}$$

$$Y_2(\beta) = B_1\cosh(\lambda\beta) + B_2\sinh(\lambda\beta) + B_3\cos(\lambda\beta) + B_4\sin(\lambda\beta) \quad \text{if } \beta \in (\beta,0] \tag{2.58}$$

where

$$\beta = \frac{z_{crack}}{L} \tag{2.59}$$

z_{crack} is crack location along the rotor, A_i and B_i $(i = 1,2,3,4)$ are arbitrary constants to be determined from the following eight boundary conditions

$$Y_1(\beta)|_{\beta=0} = 0 \tag{2.60}$$

$$Y_1''(\beta)|_{\beta=0} = 0 \tag{2.61}$$

$$Y_2(\beta)|_{\beta=1} = 0 \tag{2.62}$$

$$Y_2''(\beta)|_{\beta=1} = 0 \tag{2.63}$$

The continuity at the crack position is as follows

$$Y_1(\beta)|_\beta = Y_2(\beta)|_\beta \tag{2.64}$$

$$Y_1''(\beta)|_\beta = Y_2''(\beta)|_\beta \tag{2.65}$$

$$Y_1'''(\beta)|_\beta = Y_2'''(\beta)|_\beta \tag{2.66}$$

The compability condition due to the local flexibility is

$$Y_2'(\beta)|_\beta - Y_1'(\beta)|_\beta = c_L \frac{EI}{L}\, Y_2''(\beta)|_\beta \tag{2.67}$$

Substituting the eight of boundary conditions and writting the resulted equation in a matrix form, one obtains

$$
\left[
\begin{array}{ccccc}
1 & 0 & 1 & 0 & 0 \\
1 & 0 & -1 & 0 & 0 \\
0 & 0 & 0 & 0 & \cosh\lambda \\
0 & 0 & 0 & 0 & \cosh\lambda \\
\cosh(\lambda\beta) & \sinh(\lambda\beta) & \cos(\lambda\beta) & \sin(\lambda\beta) & -\cosh(\lambda\beta) \\
\cosh(\lambda\beta) & \sinh(\lambda\beta) & -\cos(\lambda\beta) & -\sin(\lambda\beta) & -\cosh(\lambda\beta) \\
\sinh(\lambda\beta) & \cosh(\lambda\beta) & \sin(\lambda\beta) & -\cos(\lambda\beta) & -\sinh(\lambda\beta) \\
\sinh(\lambda\beta) & \cosh(\lambda\beta) & -\sin(\lambda\beta) & \cos(\lambda\beta) & \psi\lambda\cosh(\lambda\beta)-\sinh(\lambda\beta)
\end{array}
\right.
$$

$$
\left.
\begin{array}{ccc}
0 & 0 & 0 \\
0 & 0 & 0 \\
\sinh\lambda & \cos\lambda & \sin\lambda \\
\sinh\lambda & -\cos\lambda & -\sin\lambda) \\
-\sinh(\lambda\beta) & -\cos(\lambda\beta) & -\sin(\lambda\beta) \\
-\sinh(\lambda\beta) & \cos(\lambda\beta) & \sin(\lambda\beta) \\
-\cosh(\lambda\beta) & -\sin(\lambda\beta) & \cos(\lambda\beta) \\
\psi\lambda\sinh(\lambda\beta)-\cosh(\lambda\beta) & -\psi\lambda\cosh(\lambda\beta)+\sin(\lambda\beta) & -\psi\lambda\sinh(\lambda\beta)-\cos(\lambda\beta)
\end{array}
\right]
\begin{bmatrix}
A_1 \\ A_2 \\ A_3 \\ A_4 \\ B_1 \\ B_2 \\ B_3 \\ B_4
\end{bmatrix} = 0 \tag{2.68}
$$

where

$$\psi = c_L \frac{EI}{L^3} \tag{2.69}$$

The characteristic equation can be obtained by expanding the determinants of the coefficient matrix given in Eq.(2.68) to zero. The local stiffness c_L is approximated from dimensionless local flexibility of the cracked section $\bar{c}$ which varies with the relative crack depth a/d (Dimarogonas and Papadopoulos [35]), as shown in Figure 2.34.

For $a/d = 0.1$, from Figure 2.34 we get the dimensionless local flexibility $\bar{c} \cong 0.2$, then the local stiffness c_L can be calculated from

$$\bar{c} = c_L E \left(\frac{1}{2}d\right)^3 \tag{2.70}$$

The eigenfrequencies can be obtained numerically using the secant method to solve the characteristic equation (Eq.(2.68)). After iteration, the values of $\lambda_1 = 3.1406$, $\lambda_2 = 6.2812$, and $\lambda_3 = 9.4218$ are obtained with an accuracy of 10^{-5}. The eigenfrequencies of the open cracked shaft ω_{cr} can be determined

$$\omega_{cr} = \left(\frac{\lambda_i}{L}\right)^2 \sqrt{\frac{EI}{\rho A}} \quad i = 1, 2, 3, \ldots \tag{2.71}$$

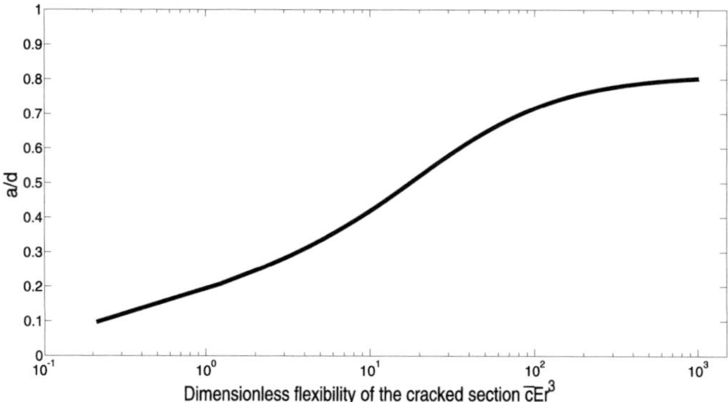

Figure 2.34: Dimensionless flexibility of the cracked section for load direction normal to crack edge

Hence

$$\omega_{cr_1} = 1\,020,305\,1\,\text{rad/s} = 162.386\,6\,\text{Hz}$$
$$\omega_{cr_2} = 4\,081,220\,5\,\text{rad/s} = 649.546\,4\,\text{Hz}$$
$$\omega_{cr_3} = 9\,182,746\,1\,\text{rad/s} = 1\,416.147\,9\,\text{Hz}$$

Finally, for comparison purpose, the first three natural frequencies of the cracked shaft are calculated using Mayes' model, local flexibility and CZM are presented in Table 2.5.

Table 2.5: Natural frequencies of uncracked shaft at rest in [rad/s]

Mode	Uncracked shaft	Cracked shaft Mayes' model	Cracked shaft local flexibility	Cracked shaft based on CZM
1	1 020.9	1 019.1	1 020.3	1 017.8
2	4 083.8	4 076.2	4 081.2	4 068.5
3	9 188.5	9 171.4	9 182.7	9 164.2

The first three natural frequencies of cracked shaft using CZM are lower than for the other open crack models due to the following reasons:

- in front of the crack tip cohesive zone element may be open partially and this may reduce the natural frequencies slightly;

- convergence of the FE solution plays an important role. Here, more continuum and cohesive elements could be used, but the calculation is time consuming.

3 Rotor with breathing transverse crack

This chapter deals with the modeling of a de Laval rotor with a transverse breathing crack. The stress intensity factor (SIF), used in fracture mechanics is valid only for open cracks. To model transition in cracks from open to close (breathings), much research has been devoted and many researchers have published a lot of studies and investigation, to study switching mechanism, i.e. from open to close. In this study, different approaches to model the breathing steering function during the rotation of a shaft, with different crack shape models that are available in the literature will be presented.

This chapter emphasisses on the dynamics of cracked de Laval rotors. First, by studying the shaft stiffness variation due to the breathing mechanism during rotation of a shaft based on linear elastic fracture mechanics (LEFM) and by adopting some reported experimental results of the crack propagation on a cracked shaft, the breathing crack shape during rotation of a shaft can be modelled. The breathing crack shape is modelled by a parabolic shape, that opens and closes due to bending stresses. It will be shown that the parabolic breathing function is considerably more general and accurate than the previously used functions in the literature [11], [47]. It can be noted that as long as the relative crack depth or crack depth ratio $a/d \leq 0.2$, the model of breathing crack parallel to crack front line or straight line may be used while the parabolic line should be used in case of deep crack, i.e. $a/d \geq 0.2$. Cohesive zone model (CZM) is applied to estimate the shaft stiffness variation on a cracked rotor. This method is valid for breathing cracks and is more simple than the other models which are based on LEFM. Furthermore, the CZM can be implemented easily in conjunction with FE models. The aim of this Chapter, is to find the actual time-periodic breathing steering functions which can be used in formulating the time-varying stiffness in the equations of motion of the cracked de Laval rotor.

3.1 Breathing crack modeling

There are many approaches reported in the literature for modeling breathing cracks in shafts. The breathing mechanism occurs in the cracked rotor when the vibration is dominated by the static deflection of the shaft. Almost all studies on cracked rotors with the assumption of weight dominance have identified little change in the form of the equations of motion of cracked rotors [21]. Figure 3.1 shows an end view of the whirling of de Laval rotor, with coordinate that describe its motion. The center of mass of the unbalanced disk is S. The point G locates the geometric center of the disk. Thus, the amount of static unbalance or the mass eccentricity is denoted $\varepsilon = \overline{GS}$, and the shaft bending deflection due to the dynamic loads is $\overline{OG}$. The stationary coordinate are O-xyz while the rotating coordinate is O-$\eta\xi\zeta$ and based on the crack direction, which rotates at the shaft rotational speed Ω. Ωt is the rotation angle, δ is the unbalance rotation angle and a is the transverse crack depth in a shaft. The rotating coordinate has an advantage of giving the synchronous whirl solution in terms of constants that are readily interpreted.

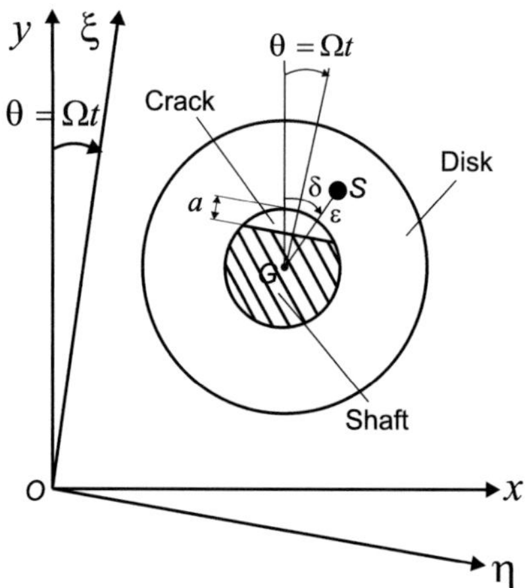

Figure 3.1: Stationary and rotating coordinate system of the cracked shaft

In general, the equations of motion of a cracked rotor with a constant rotation speed Ω can be obtained by applying Newton's Second Law to the disk and using stationary coordinates [22], [43], [44], [156]

$$\begin{bmatrix} m & 0 \\ 0 & m \end{bmatrix} \left\{ \begin{array}{c} \ddot{y} \\ \ddot{x} \end{array} \right\} + \begin{bmatrix} c & 0 \\ 0 & c \end{bmatrix} \left\{ \begin{array}{c} \dot{y} \\ \dot{x} \end{array} \right\} + \left\{ \begin{array}{c} f_{ky} \\ f_{kx} \end{array} \right\} = m\varepsilon\Omega^2 \left\{ \begin{array}{c} \cos(\theta + \delta) \\ \sin(\theta + \delta) \end{array} \right\} + \left\{ \begin{array}{c} mg \\ 0 \end{array} \right\} \quad (3.1)$$

or in vector form, the general equations of motion of a cracked rotor can be written as

$$\mathbf{M}\ddot{\mathbf{q}} + \mathbf{C}\dot{\mathbf{q}} + \mathbf{F}_k = \mathbf{F}^{ub} + \mathbf{F}^g \quad (3.2)$$

where $\mathbf{M}$ and $\mathbf{C}$ are the mass and damping matrices (including gyroscopic effects), respectively. $\mathbf{F}_k$ is the the vector of elastic force due to deformation of the shaft. $\mathbf{F}^{ub}$ and $\mathbf{F}^g$ are the centrifugal force vector of the unbalanced mass and weight force vector, respectively.

With the assumption of weight dominance, the breathing crack behaviour in the equations of motion has been expressed mostly in three forms [21], [34], [41], [99]

- The breathing crack is represented by a periodic time varying system and the governing equations of the cracked rotor are bilinear equations [41], [128], [159]

$$\left\{ \begin{array}{c} f_{k\xi} \\ f_{k\eta} \end{array} \right\} = \left[\left[\begin{array}{cc} k & 0 \\ 0 & k \end{array} \right] - f\left(t\right) \left[\begin{array}{cc} \Delta k & 0 \\ 0 & 0 \end{array} \right] \right] \left\{ \begin{array}{c} \xi \\ \eta \end{array} \right\} \tag{3.3}$$

$$\mathbf{F}_k^{rot} = \left[\mathbf{K} - f\left(t\right) \Delta \mathbf{K} \right] \mathbf{q}^{rot} \tag{3.4}$$

where $\mathbf{F}_k^{rot}$ and $\mathbf{q}^{rot}$ are vector of elastic force and vector of displacement in rotating coordinate, respectively. $f(t)$ is breathing steering function which will be discussed in the next section. From Figure 3.1, the transformation matrix between the stationary and the rotating coordinate can be directly obtained

$$\left\{ \begin{array}{c} \xi \\ \eta \end{array} \right\} = \left[\begin{array}{cc} \cos\theta & \sin\theta \\ -\sin\theta & \cos\theta \end{array} \right] \left\{ \begin{array}{c} y \\ x \end{array} \right\} \tag{3.5}$$

$$\mathbf{q}^{rot} = \mathbf{T}\left(\theta\right) \mathbf{q} \tag{3.6}$$

where $\mathbf{q}$ is the displacement in stationary coordinate. Thus, the elastic force of the shaft in the stationary coordinate is written as

$$\mathbf{F}_k = \mathbf{T}^T\left(\theta\right) \mathbf{F}_k^{rot} = \mathbf{T}^T\left(\theta\right) \left[\mathbf{K} - f\left(t\right) \Delta \mathbf{K} \right] \mathbf{T}\left(\theta\right) \mathbf{q} \tag{3.7}$$

$$\begin{aligned} \left\{ \begin{array}{c} f_{ky} \\ f_{kx} \end{array} \right\} &= \left[\left[\begin{array}{cc} k & 0 \\ 0 & k \end{array} \right] - f\left(t\right) \Delta k \left[\begin{array}{cc} \cos^2\theta & \sin\theta\cos\theta \\ \sin\theta\cos\theta & \sin^2\theta \end{array} \right] \right] \left\{ \begin{array}{c} y \\ x \end{array} \right\} \\ &= \left[\left[\begin{array}{cc} k & 0 \\ 0 & k \end{array} \right] - \frac{1}{2} f\left(t\right) \Delta k \left[\begin{array}{cc} 1+\cos 2\theta & \sin 2\theta \\ \sin 2\theta & 1-\cos 2\theta \end{array} \right] \right] \left\{ \begin{array}{c} y \\ x \end{array} \right\} \end{aligned} \tag{3.8}$$

- The breathing crack is multiplied by the stiffness matrix of the shaft directly [117], [154], [85], [107]

$$\left\{ \begin{array}{c} f_{k\xi} \\ f_{k\eta} \end{array} \right\} = \left[\left[\begin{array}{cc} k & 0 \\ 0 & k \end{array} \right] - f\left(t\right) \left[\begin{array}{cc} \Delta k_\xi & 0 \\ 0 & \Delta k_\eta \end{array} \right] \right] \left\{ \begin{array}{c} \xi \\ \eta \end{array} \right\} \tag{3.9}$$

$$\mathbf{F}_k^{rot} = \left[\mathbf{K} - f\left(t\right) \Delta \mathbf{K} \right] \mathbf{q}^{rot} \tag{3.10}$$

The elastic force of the shaft in the stationary coordinate is written as

$$\mathbf{F}_k = \mathbf{T}^T\left(\theta\right) \mathbf{F}_k^{rot} = \mathbf{T}^T\left(\theta\right) \left[\mathbf{K} - f\left(t\right) \Delta \mathbf{K} \right] \mathbf{T}\left(\theta\right) \mathbf{q} \tag{3.11}$$

$$\begin{aligned} \left\{ \begin{array}{c} f_{ky} \\ f_{kx} \end{array} \right\} &= \left[\left[\begin{array}{cc} k & 0 \\ 0 & k \end{array} \right] - f\left(t\right) \left[\begin{array}{cc} \Delta k_\xi \cos^2\theta + \Delta k_\eta \sin^2\theta & \left(\Delta k_\xi - \Delta k_\eta\right)\sin\theta\cos\theta \\ \left(\Delta k_\xi - \Delta k_\eta\right)\sin\theta\cos\theta & \Delta k_\xi \sin^2\theta + \Delta k_\eta \cos^2\theta \end{array} \right] \right] \left\{ \begin{array}{c} y \\ x \end{array} \right\} \\ &= \left[\left[\begin{array}{cc} k & 0 \\ 0 & k \end{array} \right] - \frac{1}{2} f\left(t\right) \left[\begin{array}{cc} \Delta k_1 + \Delta k_2 \cos 2\theta & \Delta k_2 \sin 2\theta \\ \Delta k_2 \sin 2\theta & \Delta k_1 + \Delta k_2 \cos 2\theta \end{array} \right] \right] \left\{ \begin{array}{c} y \\ x \end{array} \right\} \end{aligned} \tag{3.12}$$

where

$$\begin{aligned} \Delta k_1 &= \Delta k_\xi + \Delta k_\eta \tag{3.13} \\ \Delta k_2 &= \Delta k_\xi - \Delta k_\eta \tag{3.14} \end{aligned}$$

- The breathing crack is modeled by the variation in the stiffness of the cracked shaft which is expressed by a truncated cosine series. The elastic force is assumed as a function of the rotation speed Ω

$$\mathbf{F}_k = [\mathbf{K} - f(\Omega t)\,\Delta\mathbf{K}]\,\mathbf{q} \tag{3.15}$$

Sinou and Lees [137], [135] suggested $f(\Omega t)$ by a finite Fourier series with respect to Ω

$$\mathbf{F}_k = \mathbf{F}_{k0} + \sum_{k=1}^{m}\{F_{ck}\cos\Omega t + F_{sk}\sin\Omega t\} \tag{3.16}$$

Papadopoulos and Dimarogonas [102], Pennacchi et al. [105] expressed the stiffness as a truncated cosine series

$$\mathbf{F}_k = \mathbf{K}(\Omega t)\,\mathbf{q} = \{\mathbf{K}_0 + \mathbf{K}_1\cos\Omega t + \mathbf{K}_2\cos 2\Omega t + \mathbf{K}_3\cos 3\Omega t + \mathbf{K}_4\cos 4\Omega t\}\,\mathbf{q} \tag{3.17}$$

where $\mathbf{K}_1$, $\mathbf{K}_2$, $\mathbf{K}_3$, and $\mathbf{K}_4$ are the stiffness matrices which depend on the uncracked condition, a fully open crack and a half-open and half-closed crack, respectively. These suggested elastic forces are suitable only in the case of weight dominance.

3.1.1 Breathing steering functions

Three breathing crack models and open crack model are introduced and compared [88]. The crack may remain open throughout a full shaft revolution if the unbalance is very high. Only the open portions of the crack change the stiffness of rotor and hence the stiffness will be a function of the orientation of the rotor. In this case the crack does not breath and it is assumed to be open all the time, and breathing steering function can be written as

$$f(\Omega t) = 1 \tag{3.18}$$

If the crack remains open, the rotor is then local asymmetric and this condition can lead to instability problems. If a cracked shaft rotates slowly under the load of its own weight, then the crack will open and close once per revolution, it is called breathing crack. Mayes and Davies [89] proposed a breathing crack model in which the opening and closing of the crack was described by cosine function and intended for deep cracks. The breathing steering function [89] is defined as

$$f(\Omega t) = \frac{1}{2}(1 + \cos(\Omega t)) \tag{3.19}$$

The simplest model of an opening and closing crack is the hinge model proposed by Gasch [40]. In this model, the crack is assumed to change from its closed to open state abruptly as the shaft rotates. Thus, in the hinge model the breathing steering function is defined as

$$f(\Omega t) = \begin{cases} 1 & -\frac{\pi}{2} \le \Omega t < \frac{\pi}{2} & \text{crack closes;} \\ 0 & \frac{\pi}{2} \le \Omega t < \frac{3\pi}{2} & \text{crack opens} \end{cases} \tag{3.20}$$

When the crack is closed, shaft stiffness is equal to an undamaged shaft. The Gasch's model is a step function, presumably for shallow depth cracks. Step function of Gasch's model can be approximated mathematically as the finite truncated Fourier series [88]

$$f\left(\Omega t\right) = \frac{2}{\pi}\left(\frac{\pi}{4} + \cos\left(\Omega t\right) - \frac{1}{3}\cos\left(3\Omega t\right) + \frac{1}{5}\cos\left(5\Omega t\right) - \frac{1}{7}\cos\left(7\Omega t\right) + \cdots\right) \quad (3.21)$$

The hinge model suffers from the defect that there is no direct relationship between the shaft stiffness and the depth of crack. The Mayes' model does provide a method to determine the reduction in stiffness from crack depth [106]. Furthermore, for deeper cracks, the Mayes' model is better than the Gasch's model on the unstable zone, because the Mayes' model includes the cross flexibility, which the hinge model ignores [41].

The other breathing steering function is proposed by Yang et al. [158]. They suggested breathing steering function where the relative crack depth a/d (a is crack depth and d is diameter of shaft) influence is strongly accounted. Its expression is as follows,

$$f\left(\Omega t\right) = \left(\frac{1 + \cos\left(\Omega t\right)}{2}\right)^{\frac{2a}{d}} \quad (3.22)$$

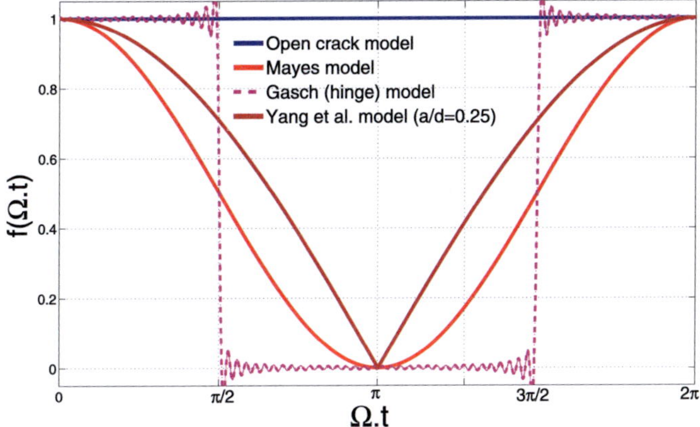

Figure 3.2: Open crack model and various breathing function models

In Figure 3.2, a comparison between the three breathing steering function models and the open crack model. Comparing Mayes' and Yang's models, have the same values for the case of deep cracks, but for the shallow cracks Yang proposed model becomes apart and deviates further from Gasch model as shown in Figure 3.3.

3.1.2 Breathing crack shapes

Cracks perpendicular to the shaft axis are known as transverse cracks. Most of the research focuses on the detection of these most dangerous cracks. Due to shaft self-weight

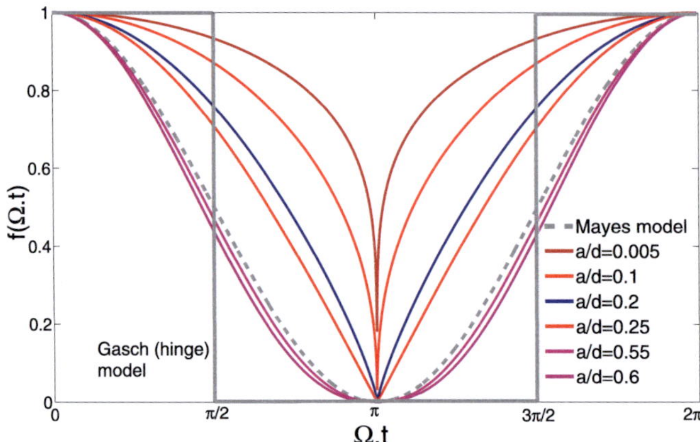

Figure 3.3: Comparison between open, Mayes' and Yang's crack model for different relative crack depth a/d

and the rotation of the rotor, the crack opens and closes during a complete revolution of the rotor and the stiffness of the shaft varies. Cracks which open when the affected part of the material is subjected to tensile stresses and close when the stress is reversed are known as breathing cracks. Usually, shaft cracks breathe when crack sizes are small, running speeds are low and radial forces are large [47]. Most theoretical research efforts concentrate on transverse breathing cracks due to their direct practical relevance. Most of the published literature on breathing crack model has considered some assumptions in modelling the breathing crack during rotation of shaft. Different assumptions have been used in the literature for modelling the transverse crack in rotating shafts e.g. a crack front line that is perpendicular to the crack front extensively addressed by Darpe et al. [30]. This breathing crack shape was also used by Jun et al. [66], and Sinou and Lees [136] (Figure 3.4). An elliptical shape has been proposed by Shih [131], Bachschmid et al. [11] (Figure 3.5).

The equivalent crack model is suggested by Lees and Friswell [77]. They calculate the area of open crack at each orientation of the rotor. The compliance functions in the two orthogonal directions are then evaluated by assuming equivalence with the effective area. Figure 3.6 shows graphically the open and closed portions for a particular crack depth and orientation their model.

Al-Shudeifat and Butcher [2] represent a breathing function for transverse breathing crack. They demonstrate, as the shaft starts to rotate, that the locations of the centroid and the neutral axis of the crack element are changed with time during rotation as shown in Figure 3.7. The tension stress field exists below the neutral axis which tends to keep the crack open. The compression stress field that exists above the neutral axis tends to compress the crack to be closed as represented in the Figure 3.7.

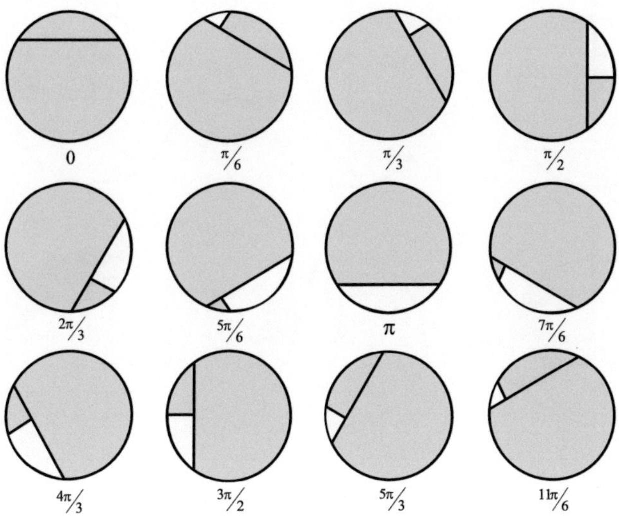

Figure 3.4: Crack state variations during rotation: perpendicular to crack front line model [30], [66], [136]

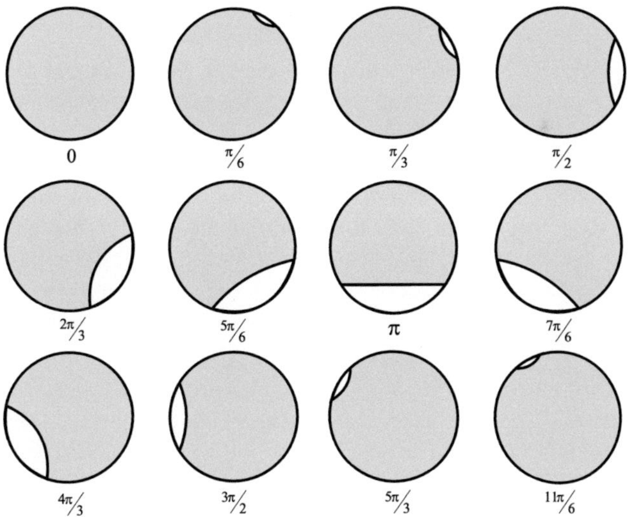

Figure 3.5: Crack state variations during rotation: elliptical crack model [131], [11]

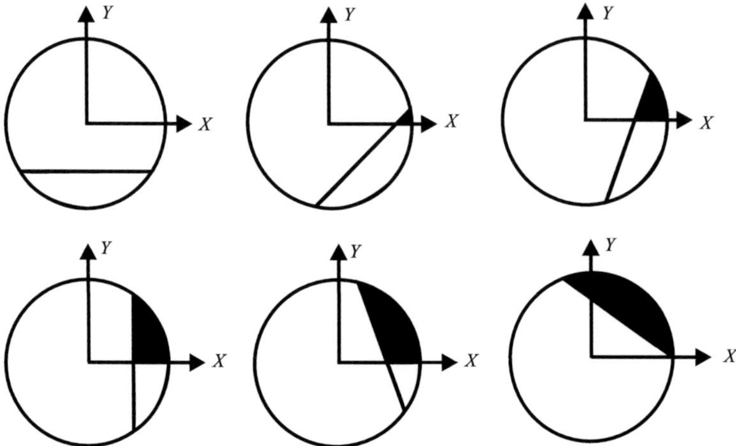

Figure 3.6: Crack state variations during rotation suggested by Lees and Friswell [77]

Another model is introduced by Liong and Proppe [80], [83] where the crack opens and closes parallel to the crack front line as long as the crack depth is small (Figure 3.8). This breathing crack model is based on an elliptical crack model. Details of this model will be discussed in the next section.

Implementation of the CZM to study breathing crack using FE model is presented in Section 2.7. As can be seen there are some relevant differences with respect to the model used by Darpe et al. (Figure 3.4) and used by Shih and Chen (Figure 3.5). In the FE model, the crack opens more slowly at the beginning, but increases its opening speed at 60° and at 90° it is more open and at 150° it is already completely open. The limitation these results are due to quasi-static condition. One revolution of the shaft is divided in angular steps of 30°, a bending load is applied and the calculation is repeated for all different angular positions of the cracked shaft specimen, which at each step the open crack condition is observed. Almost identical results can be found in Figure 3.9 [10] presented by using FE model.

Bachschmid et al. [11], [12] reported that crack, which are propagating due to bending stresses in rotating shafts, have often an elliptical shape as long as the depth is small. They presented their results in Figure 3.10 and 3.11 to illustrate the shapes of cracks, which were obtained by applying static bending moment to rotating shaft specimen, respectively in a very early propagation state and after the crack has propagated to a consistent depth. In both shafts crack initiation was obtained by means of a small cut (notch) visible in both figures. Furthermore, Lorentzen et al. [86] developed a theoretical method for calculating the SIF for a cracked shaft subjected to a constant moment load. They also conducted

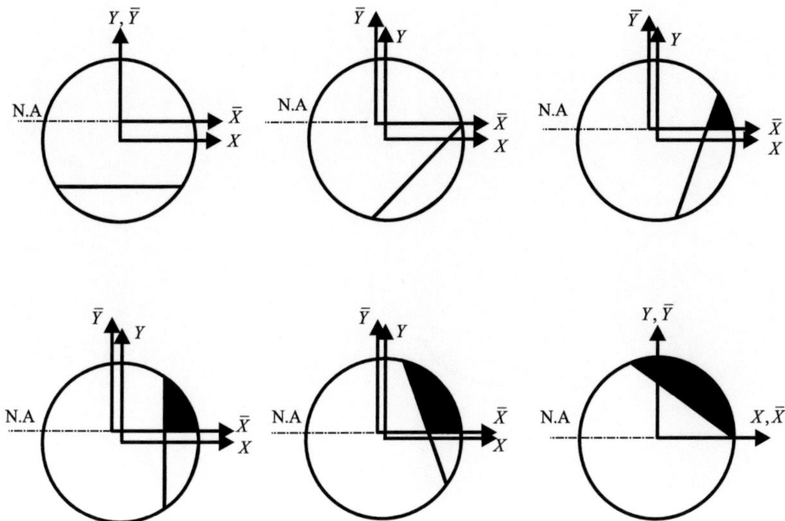

Figure 3.7: Crack state variations during rotation represented by Al-Shudeifat and Butcher [2]

some experiments using 2D and 3D photoelasticity to predict the critical crack depth, crack shape and load configuration. Their experimental results are shown in Figure 3.12 and showed good agreement with the experimental results reported by Bachschmid et al. [11], [12]. Härkegard et al. [57] investigated growth of naturally initiated fatigue cracks in gas turbine rotor. They used cylindrical test specimen with semi elliptic and proposed the crack growth rates for the cracked rotor.

Based on FE and MBS (Chapter 6) and some reported experimental results (Figure 3.12) and in order to model more realistically the breathing crack mechanism during rotation of rotor, the breathing crack shape is modelled by a crack closure parabolic shape, that opens and closes due to bending stresses as shown in Figure 3.13. It will be shown that the crack closure parabolic line model introduced in this section is considerably more general and accurate than the previously used functions in the literature. It can be noted that as long as the relative crack depth is small ($a/d \leq 0.2$), the model of breathing crack parallel to crack front line or crack closure straight line (Figure 3.8) may be used while the crack closure parabolic line should be used in case of deep crack ($a/d \geq 0.2$).

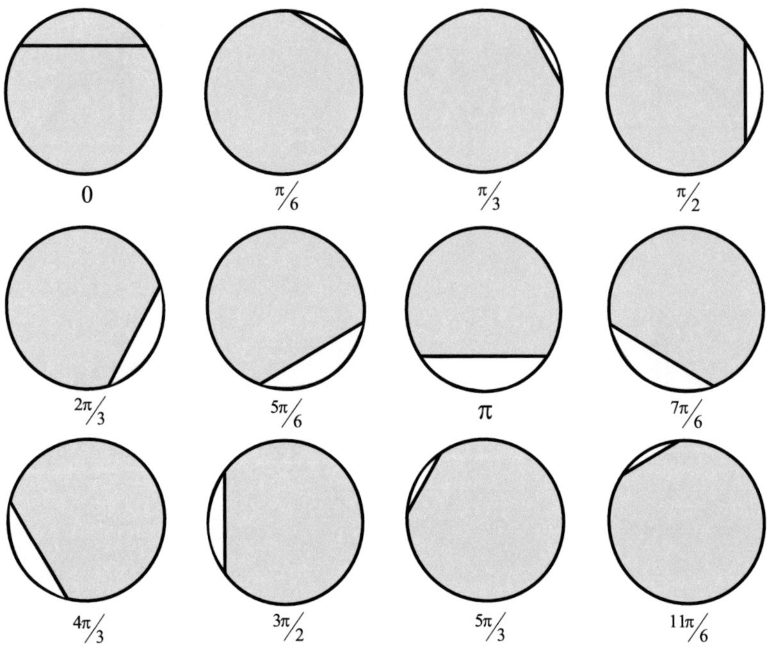

Figure 3.8: Crack state variations during rotation: parallel to crack front line model [80], [83]

3.2 Dynamics of cracked rotors

3.2.1 Mathematical formulation

Refering to the end view of the de Laval rotor shown in Figure 3.1, the shaft has a bending stiffness k carrying an unbalanced disk at its midspan. The disk has a mass of m and air drag on the rotating disk and shaft is approximated by a viscous damping coefficient of c. The shaft is assumed to be with negligible mass and have a transverse crack of crack depth a at its midspan. The shaft is supported radially by rigid bearings and the crack opens in the direction of ξ as shown in Figure 3.1.

The equations of motion can be expressed in rotating coordinates as

$$\left\{ \begin{array}{c} y \\ x \end{array} \right\} = \left(\begin{array}{cc} \cos\theta & -\sin\theta \\ \sin\theta & \cos\theta \end{array} \right) \left\{ \begin{array}{c} \xi \\ \eta \end{array} \right\} \tag{3.23}$$

In stationary coordinates

$$r = y + jx = le^{j(\theta+\delta)} \tag{3.24}$$

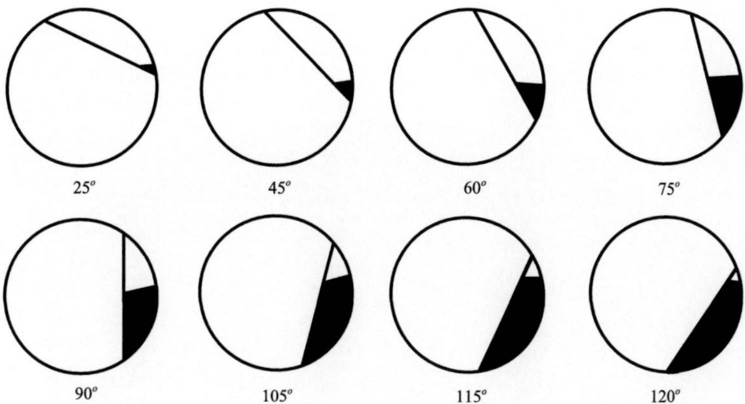

Figure 3.9: Crack state variations during rotation reported by Bachschmid et al. [10]

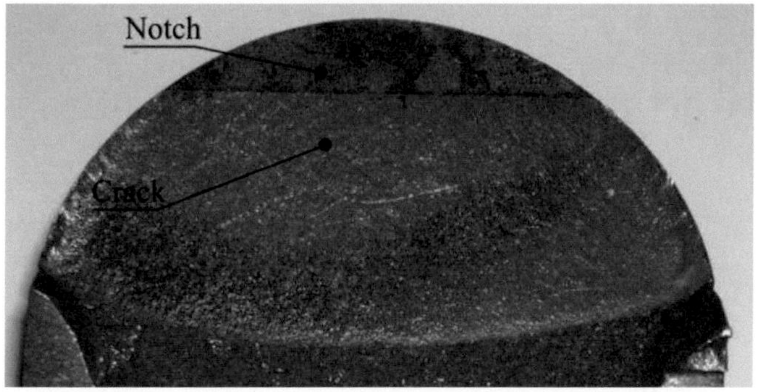

Figure 3.10: Crack in early propagation state [11]

and in rotating coordinates

$$\rho = \xi + j\eta = le^{j\delta} \tag{3.25}$$

where

$$
\begin{aligned}
r &= le^{j\theta}e^{j\delta} = \rho e^{j\theta} & (3.26)\\
\dot{r} &= (\dot{\rho} + j\dot{\theta}\rho)e^{j\theta} & (3.27)\\
\ddot{r} &= (\ddot{\rho} + 2j\dot{\theta}\dot{\rho} - \dot{\theta}^2\rho)e^{j\theta} & (3.28)
\end{aligned}
$$

An assumption of δ constant yields synchronous whirl, hence

$$\theta = \Omega t \tag{3.29}$$

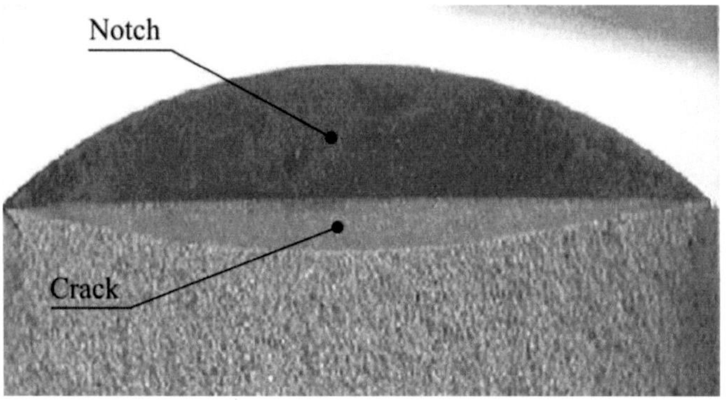

Figure 3.11: Deep propagated crack [11]

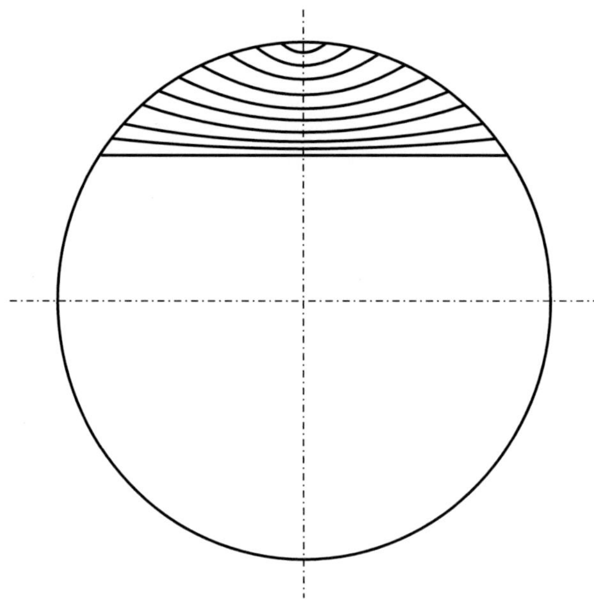

Figure 3.12: Experimentally determined crack propagation patterns [86], [11]

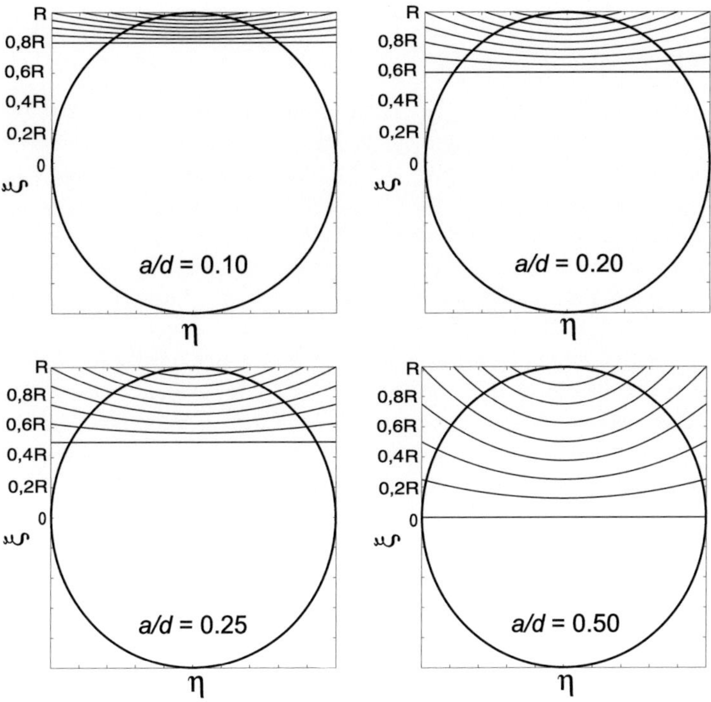

Figure 3.13: Crack closure parabolic line model for various relative crack depths a/d

Thus

$$m\ddot{r} + c\dot{r} + kr = m\varepsilon\dot{\theta}^2 e^{j(\theta+\delta)} + mge^{j\pi} \tag{3.30}$$

$$m(\ddot{\rho} + 2j\dot{\theta}\dot{\rho} - \dot{\theta}^2\rho)e^{j\theta} + c(\dot{\rho} + j\dot{\theta}\rho) + k\rho e^{j\theta} = m\varepsilon\dot{\theta}^2 e^{j\theta}e^{j\delta} + mge^{j\pi} \tag{3.31}$$

Real and imaginary parts are

$$m\ddot{\xi} - 2m\dot{\theta}\dot{\eta} - m\dot{\theta}^2\xi + c\dot{\xi} - c\dot{\theta}\eta + k_\xi\xi = m\varepsilon\dot{\theta}^2\cos\delta - mge^{-j\theta} \tag{3.32}$$

$$m\ddot{\eta} + 2m\dot{\theta}\dot{\xi} - m\dot{\theta}^2\eta + c\dot{\eta} + c\dot{\theta}\xi + k_\eta\eta = m\varepsilon\dot{\theta}^2\sin\delta - mge^{-j\theta} \tag{3.33}$$

The equations of motion are obtained as

$$m(\ddot{\xi} - 2\dot{\theta}\dot{\eta} - \dot{\theta}^2\xi) + c(\dot{\xi} - \dot{\theta}\eta) + k_\xi\xi = m\varepsilon\dot{\theta}^2\cos\delta - mg\cos\theta \tag{3.34}$$

$$m(\ddot{\eta} + 2\dot{\theta}\dot{\xi} - \dot{\theta}^2\eta) + c(\dot{\eta} + \dot{\theta}\xi) + k_\eta\eta = m\varepsilon\dot{\theta}^2\sin\delta + mg\sin\theta \tag{3.35}$$

3.2.2 Equations of motion of the de Laval rotor with breathing crack

The typical breathing behaviour in the form of continuous change in the stiffness of the cracked rotor is due to the opening and closing of the crack under the effect of gravity acting on the horizontal rotor. Due to the presence of gravity, the upper portion of the cracked rotor at the start of the rotation is under compression and the crack is closed. As the rotor continues to rotate and the gravity direction being constant, the upper part now comes in the lower tensile region causing the crack to open. The process repeats and a periodic crack opening and closing phenomenon called crack breathing results.

In most works the reduction of stiffness due to crack opening is considered along the weaker axis only, but it is well known that the stiffness also reduces along the stronger axis as the crack propagates. Many investigators used a step function to express the change of stiffness, assuming a sudden occurrence of crack opening and closing or a function $(1+\cos\theta)/2$, $\theta = \Omega t$ being the angle between the crack and the force due to gravity, in order to allow partial crack opening or closing. In order to analyse the effects of a crack in the shaft the equations of motion should be formulated with consideration of the stiffness modification. In this section, the cross-coupled stiffnesses as well as the direct stiffnesses are estimated by fracture mechanics concepts in order to consider the partial opening and closing behaviour of a breathing crack.

$$m(\ddot{\xi} - 2\dot{\theta}\dot{\eta} - \dot{\theta}^2\xi) + c(\dot{\xi} - \dot{\theta}\eta) + k_\xi\xi + k_{\xi\eta}\eta = m\varepsilon\dot{\theta}^2\cos\delta - mg\cos\theta \tag{3.36}$$

$$m(\ddot{\eta} + 2\dot{\theta}\dot{\xi} - \dot{\theta}^2\eta) + c(\dot{\eta} + \dot{\theta}\xi) + k_\eta\eta + k_{\eta\xi}\xi = m\varepsilon\dot{\theta}^2\sin\delta + mg\sin\theta \tag{3.37}$$

When a well balanced horizontal rotor with a transverse crack rotates, the crack breathes, opening and closing once per revolution due to the gravitational force. Numerical simulation results in [66] showed that the direct stiffnesses tend to decrease as the crack starts opening and tend to increase as the crack starts closing, whereas the cross-coupled stiffnesses vanish at the completely open and closed states and take maximum values somewhere in between the completely opened and closed states.

In the present study, the following assumptions are made:

1. The location of the crack is assumed to be known at the mid-span of the shaft, i.e. at the maximum deflection point.

2. Only a single transverse crack has been considered.

3. Plane strain condition is considered at the crack front due to the geometry constraint.

4. The shaft is rigidly supported at both ends (flexible rotor).

5. The structural damping is constant.

6. The shaft has uniform cross-section A, length $L = 1.0$ m and diameter $d = 0.08$ m of the shaft.

7. The material of the shaft is considered to be homogeneous and isotropic. Modulus of elasticity E, Poisson's ratio ν and mass density ρ are 210 GPa, 0.3 and 7 850 kg/m^3, respectively. The power law strain hardening rule is used for plasticity with yield strength and ultimate strength of material 250 MPa and 400 MPa, respectively.

3.3 Stiffness estimation based on linear elastic fracture mechanics

The breathing crack model proposed by Jun et al. [66] and Darpe et al. [30] is discussed. In the bent shaft while rotating, the forces Q_ξ and Q_η acting along the ξ and η axes on the cross-section containing the crack as shown in Figure 3.14 induce deflections of the solid shaft. The additional deflections due to the crack are estimated by using fracture mechanics concepts. When the forces Q_ξ and Q_η exist, the stress may have different values along the crack front. Hence the SIF is expressed as a function of crack width w (Figure 3.15).

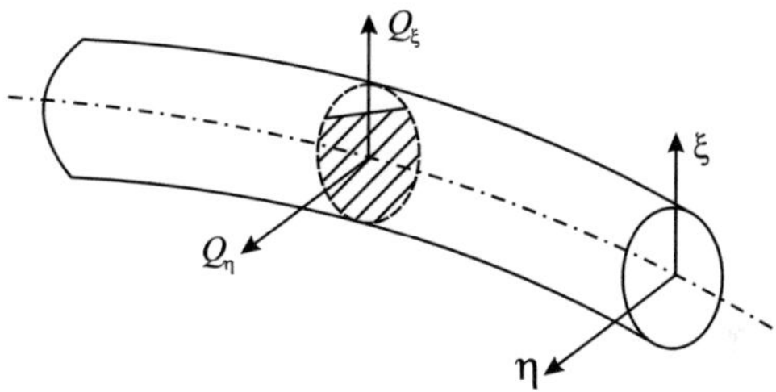

Figure 3.14: Model of shaft with a transverse crack

Since the longitudinal (z-axis) length of the shaft is very large compared to α', the SIF $K_{Q\xi}^I$ due to the forces Q_ξ (the superscript I denotes the mode-I in fracture mechanics) is approximately equal to the value for the crack under pure bending. The effect of shear stress is negligible.

The SIF $K_{Q\xi}^I$ is then given by

$$K_{Q\xi}^I = \sigma_\xi\left(w\right)\sqrt{\pi\alpha}F\left(\frac{\alpha}{\alpha'}\right) \tag{3.38}$$

$$\sigma_\xi\left(w\right) = \frac{\left(\frac{1}{4}Q_\xi L\right)\frac{1}{2}\alpha'}{I} \tag{3.39}$$

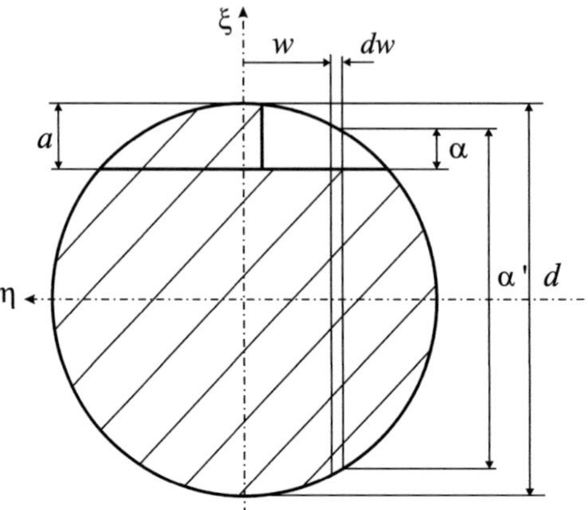

Figure 3.15: Geometry of the cracked shaft

where

$$I = \frac{\pi}{64}d^4 \tag{3.40}$$

$$F\left(\frac{\alpha}{\alpha'}\right) = \sqrt{\frac{2\alpha'}{\pi\alpha}\tan\left(\frac{\pi\alpha}{2\alpha'}\right)} \frac{0.923 + 0.199\left[1 - \sin\left(\frac{\pi\alpha}{2\alpha'}\right)\right]^4}{\cos\left(\frac{\pi\alpha}{2\alpha'}\right)} \tag{3.41}$$

$$\alpha' = \sqrt{d^2 - (2w)^2} \tag{3.42}$$

For the transverse force Q_η acting on the same strip, the SIF $K_{Q\eta}^I$ due to Q_η is considered. In a similar manner to the case of $K_{Q\xi}^I$

$$K_{Q\eta}^I = \sigma_\eta(w)\sqrt{\pi\alpha}F\left(\frac{\alpha}{\alpha'}\right) \tag{3.43}$$

$$\sigma_\eta(w) = \frac{\left(\frac{1}{4}Q_\eta L\right)w}{I} \tag{3.44}$$

$$F'\left(\frac{\alpha}{\alpha'}\right) = \sqrt{\frac{2\alpha'}{\pi\alpha}\tan\left(\frac{\pi\alpha}{2\alpha'}\right)} \frac{0.752 + 2.02\left(\frac{\alpha}{\alpha'}\right) + 0.37\left[1 - \sin\left(\frac{\pi\alpha}{2\alpha'}\right)\right]^3}{\cos\left(\frac{\pi\alpha}{2\alpha'}\right)} \tag{3.45}$$

The total SIF K^I is given by

$$K^I = K_{Q\xi}^I + K_{Q\eta}^I \tag{3.46}$$

where the strain energy density function is given by

$$
\begin{aligned}
J(\alpha) &= \frac{1}{E}\left(K^{I}_{Q\xi} + K^{I}_{Q\eta}\right)^2 \\
&= \frac{1}{E}\left(\sigma_\xi^2 \pi\alpha F\left(\frac{\alpha}{\alpha'}\right)^2 + \sigma_\eta^2 \pi\alpha F'\left(\frac{\alpha}{\alpha'}\right)^2\right) + \\
&\quad \frac{1}{E}\left(2\sigma_\xi\sigma_\eta\pi\alpha F\left(\frac{\alpha}{\alpha'}\right) F'\left(\frac{\alpha}{\alpha'}\right)\right) \\
&= \frac{64L^2\alpha'^2\alpha}{E\pi d^8}Q_\xi^2 F\left(\frac{\alpha}{\alpha'}\right)^2 + \frac{256L^2 w^2\alpha}{E\pi d^8}Q_\eta^2 F'\left(\frac{\alpha}{\alpha'}\right)^2 + \\
&\quad \frac{256L^2\alpha' w\alpha}{E\pi d^8}Q_\xi Q_\eta F\left(\frac{\alpha}{\alpha'}\right) F'\left(\frac{\alpha}{\alpha'}\right)
\end{aligned}
\tag{3.47}
$$

The additional deflection $\hat{u}_i$, due to the crack is given by

$$
\hat{u}_i = \frac{\partial}{\partial Q_i}\left[\int J(\alpha)\,\mathrm{d}\alpha\right]
\tag{3.48}
$$

then

$$
\begin{aligned}
\hat{u}_\xi &= \frac{\partial}{\partial Q_\xi}\left[\int J(\alpha)\,\mathrm{d}\alpha\right] \\
&= \int\left[\frac{128L^2\alpha'^2\alpha}{E\pi d^8}Q_\xi F\left(\frac{\alpha}{\alpha'}\right)^2 + \frac{256L^2\alpha' w\alpha}{E\pi d^8}Q_\eta F\left(\frac{\alpha}{\alpha'}\right) F'\left(\frac{\alpha}{\alpha'}\right)\right]\mathrm{d}\alpha
\end{aligned}
\tag{3.49}
$$

$$
\begin{aligned}
\hat{u}_\eta &= \frac{\partial}{\partial Q_\eta}\left[\int J(\alpha)\,\mathrm{d}\alpha\right] \\
&= \int\left[\frac{512L^2 w^2\alpha}{E\pi d^8}Q_\eta F'\left(\frac{\alpha}{\alpha'}\right)^2 + \frac{256L^2\alpha' w\alpha}{E\pi d^8}Q_\xi F\left(\frac{\alpha}{\alpha'}\right) F'\left(\frac{\alpha}{\alpha'}\right)\right]\mathrm{d}\alpha
\end{aligned}
\tag{3.50}
$$

The flexibility due to the crack is now defined as

$$
g_i = \frac{\partial \hat{u}_i}{\partial Q_i}
\tag{3.51}
$$

then

$$
g_\xi = \frac{\partial \hat{u}_\xi}{\partial Q_\xi} = \iint \frac{128L^2\alpha'^2\alpha}{E\pi d^8}F\left(\frac{\alpha}{\alpha'}\right)^2\,\mathrm{d}\alpha\,\mathrm{d}w
\tag{3.52}
$$

$$
g_\eta = \frac{\partial \hat{u}_\eta}{\partial Q_\eta} = \iint \frac{512L^2 w^2\alpha}{E\pi d^8}F'\left(\frac{\alpha}{\alpha'}\right)^2\,\mathrm{d}\alpha\,\mathrm{d}w
\tag{3.53}
$$

$$
g_{\xi\eta} = \frac{\partial \hat{u}_\xi}{\partial Q_\eta} = \iint \frac{256L^2\alpha' w\alpha}{E\pi d^8}F\left(\frac{\alpha}{\alpha'}\right) F'\left(\frac{\alpha}{\alpha'}\right)\,\mathrm{d}\alpha\,\mathrm{d}w = g_{\eta\xi} = \frac{\partial \hat{u}_\eta}{\partial Q_\xi}
\tag{3.54}
$$

Adding flexibility of uncracked shaft to additional flexibility due to the crack, the following flexibility coefficients are obtained, which for completely open crack, the area integration is explicitly performed by

$$g_\xi = \frac{L^3}{48EI} + \int_{-\sqrt{a(d-a)}}^{\sqrt{a(d-a)}} \int_0^{a-\left(\frac{d}{2}-\sqrt{\left(\frac{d}{2}\right)^2-w^2}\right)} \frac{128L^2\alpha'^2\alpha}{E\pi d^8} F\left(\frac{\alpha}{\alpha'}\right)^2 d\alpha\,dw \tag{3.55}$$

$$g_\eta = \frac{L^3}{48EI} + \int_{-\sqrt{a(d-a)}}^{\sqrt{a(d-a)}} \int_0^{a-\left(\frac{d}{2}-\sqrt{\left(\frac{d}{2}\right)^2-w^2}\right)} \frac{512L^2w^2\alpha}{E\pi d^8} F'\left(\frac{\alpha}{\alpha'}\right)^2 d\alpha\,dw \tag{3.56}$$

$$g_{\xi\eta} = g_{\eta\xi} = \int_{-\sqrt{a(d-a)}}^{\sqrt{a(d-a)}} \int_0^{a-\left(\frac{d}{2}-\sqrt{\left(\frac{d}{2}\right)^2-w^2}\right)} \frac{256L^2\alpha'w\alpha}{E\pi d^8} F\left(\frac{\alpha}{\alpha'}\right) F'\left(\frac{\alpha}{\alpha'}\right) d\alpha\,dw \tag{3.57}$$

By adding the deflection without the crack to the deflection due to the crack, the total deflection can be written as

$$\Delta^\xi = Q_\xi g_\xi + Q_\eta g_{\xi\eta} = \xi \tag{3.58}$$

$$\Delta^\eta = Q_\xi g_{\eta\xi} + Q_\eta g_\eta = \eta \tag{3.59}$$

or

$$Q_\xi = \frac{g_\eta \xi - g_{\xi\eta} \eta}{g_\xi g_\eta - g_{\xi\eta}^2} \tag{3.60}$$

$$Q_\eta = \frac{g_\xi \eta - g_{\xi\eta} \xi}{g_\xi g_\eta - g_{\xi\eta}^2} \tag{3.61}$$

Using the above flexibility values, the following stiffness values are obtained

$$k_\xi = \frac{g_\eta}{g_\xi g_\eta - g_{\xi\eta}^2} \tag{3.62}$$

$$k_\eta = \frac{g_\xi}{g_\xi g_\eta - g_{\xi\eta}^2} \tag{3.63}$$

$$k_{\xi\eta} = k_{\eta\xi} = \frac{-g_{\xi\eta}}{g_\xi g_\eta - g_{\xi\eta}^2} \tag{3.64}$$

3.4 Stiffness estimation based on cohesive zone model

As discussed in the section 3.1, the crack shape is modelled by either straight or parabolic line, and the crack opens and closes due to bending stresses in rotating shafts as shown in Figure 3.13. As long as the relative crack depth is small ($a/d \leq 0.2$), the model of a breathing crack parallel to the crack front line or crack closure straight line (Figure 3.8) may be used. If the crack depth is small, the crack closure line follows very closely a straight line while for larger crack depths the crack closure line becomes more curved (Figure 3.13). This Section is dedicated to provide the crack closure straight line for shallow cracks. It is known from the literature, that the existence of a transverse crack in a shaft induces a local compliance that differs in each direction. A cracked shaft under two bending loads acting in the vertical direction and the horizontal direction is shown in Figure 3.16. As the shaft rotates, the crack takes on different angular positions at the cracked section (Figure 3.17). From the transverse forces Q_ξ, Q_η acting on the same strip, the nominal stresses are given by

$$\sigma_\xi = \frac{(\frac{1}{4}Q_\xi L)(\frac{1}{2}d - a + \alpha)}{\frac{\pi}{64}d^4} = \frac{16Q_\xi L(\frac{1}{2}d - a + \alpha)}{\pi d^4} \tag{3.65}$$

$$\sigma_\eta = \frac{(\frac{1}{4}Q_\eta L)(\frac{1}{2}w)}{\frac{\pi}{64}d^4} = \frac{16Q_\eta L\sqrt{(d - a + \alpha)(a - \alpha)}}{\pi d^4} \tag{3.66}$$

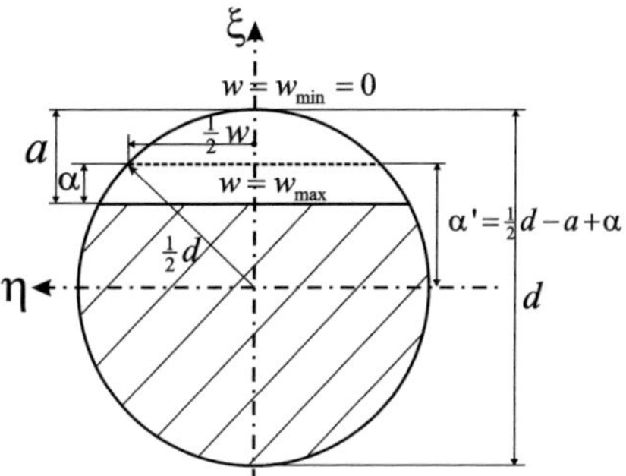

Figure 3.16: Geometry of the crack closure straight line model

In order to obtain the stiffness variation, the concept of crack closure line is applied. The crack closure line is a line parallel to the crack edge. It separates the open and closed parts of the crack as shown in Figure 3.17. It is assumed that if a well balanced horizontal rotor with a transverse crack rotates, the crack breathes, opening and closing once per revolution due to the gravitational force. Moreover, when unbalance exists in a cracked rotor, the crack may breathe, remain completely open or closed, depending on the magnitude and phase of the unbalance and the crack possible always opens and one can use the open crack model [66].

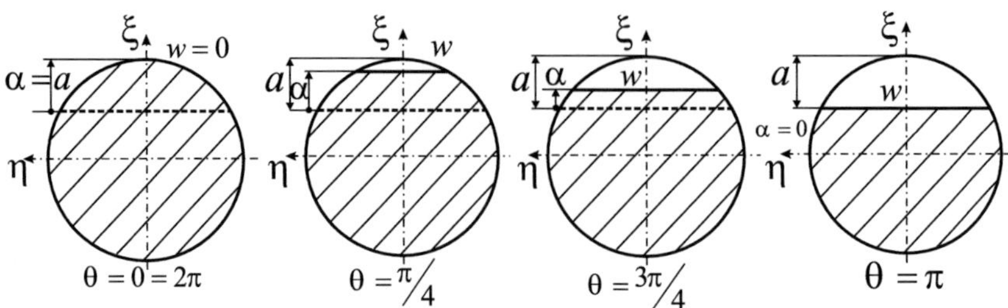

Figure 3.17: Breathing condition under rotation due to bending moment of gravity

In the bent shaft while rotating, the forces Q_ξ and Q_η acting along the ξ and η axes on the cross-section containing the crack as shown in 3.14, induce deflections of the solid shaft. The additional deflections due to the crack are estimated using the CZM. When the forces Q_ξ and Q_η exist, the stress may have different values along the crack front. From Castigliano's theorem $\hat{u} = \partial U/\partial Q$, where U is the total strain energy, that is $U = U_0 + U_c$, here U_0 and U_c are the strain energy of the uncracked shaft element and the strain energy due to the crack, respectively.

Paris equation [35] gives the additional deflection, due to a crack depth a in the i-th direction, as

$$\hat{u}_i = \frac{\partial}{\partial Q_i} \int_0^a J(\alpha)\, d\alpha \tag{3.67}$$

where $J(\alpha)$ is the strain energy density function and Q_i the corresponding load. Using Castigliano's theorem, the additional deflection $\hat{u}_i$, due to the crack in the i-th direction can be determined

$$\hat{u}_i = \frac{\partial}{\partial Q_i} \int_0^a G_I(\alpha)\, d\alpha = \frac{\partial}{\partial Q_i} \left[\int_0^a \left(\int_0^{\delta_1} \sigma_n\, d\delta \right) d\alpha \right] \tag{3.68}$$

where $G_I(\alpha)$ is the strain energy density function and δ_1 the elastic limit of cohesive element.

In the elastic region, one has

$$\sigma_n = \left(1 + \sqrt{3}\chi_{eff}\right) \frac{2E}{3}\delta \tag{3.69}$$

then

$$\hat{u}_i = \frac{\partial}{\partial Q_i} \left[\int_0^a \left(\int_0^{\delta_1} \left(1 + \sqrt{3}\chi_{eff}\right) \frac{2E}{3}\delta\, d\delta \right) d\alpha \right] \tag{3.70}$$

The additional deflection due to the crack can be formed

$$
\begin{aligned}
\hat{u}_i &= \frac{\partial}{\partial Q_i}\left[\int_0^a \left(\int_0^{\delta_1}\left(1+\frac{\sqrt{3}}{2}\left(\frac{1+r_\sigma}{\sqrt{3}\,(1-r_\sigma)}-\frac{2}{3}\ln\left(\frac{S\sigma_Y}{CE}\right)\right)\right)\frac{2E}{3}\delta\,\mathrm{d}\delta\right)\mathrm{d}\alpha\right] \\
&= \frac{\partial}{\partial Q_i}\left[\int_0^a \frac{E\delta_1^2}{3}\left(1+\frac{1}{2}\left(\frac{1+r_\sigma}{1-r_\sigma}-\frac{2}{\sqrt{3}}\ln\left(\frac{S\sigma_Y}{CE}\right)\right)\right)\mathrm{d}\alpha\right]
\end{aligned}
\tag{3.71}
$$

where the principal stress are defined by

$$
\sigma_{1,2}=\frac{\sigma_x+\sigma_y}{2}\pm\sqrt{\left(\frac{\sigma_x+\sigma_y}{2}\right)^2+\tau_{xy}^2}
\tag{3.72}
$$

$$
\tau_{xy}=0
\tag{3.73}
$$

and

$$
\sigma_1=\sigma_\eta
\tag{3.74}
$$

$$
\sigma_2=\sigma_\xi \quad (\sigma_1<\sigma_2)
\tag{3.75}
$$

Then the stress ratio is

$$
r_\sigma=\frac{\sigma_1}{\sigma_2}\leq 1
\tag{3.76}
$$

The stress ratio can be written as

$$
\begin{aligned}
r_\sigma &= \frac{\sigma_1}{\sigma_2}=\frac{\sigma_\eta}{\sigma_\xi} \\
&= \frac{\sqrt{(d-a+\alpha)\,(a-\alpha)}\,Q_\eta}{\frac{1}{2}d-a+\alpha}\frac{Q_\eta}{Q_\xi}=\frac{Q_\eta p}{Q_\xi q}
\end{aligned}
\tag{3.77}
$$

where

$$
p=\sqrt{(d-a+\alpha)\,(a-\alpha)}
\tag{3.78}
$$

$$
q=\frac{1}{2}d-a+\alpha
\tag{3.79}
$$

Therefore, the additional deflection is given by

$$
\hat{u}_i=\frac{\partial}{\partial Q_i}\left[\int_0^a \frac{E\delta_1^2}{3}\left(1+\frac{1}{2}\frac{Q_\xi q+Q_\eta p}{Q_\xi q-Q_\eta p}-\frac{1}{\sqrt{3}}\ln\left(\frac{S\sigma_Y}{CE}\right)\right)\mathrm{d}\alpha\right]
\tag{3.80}
$$

For ξ and η direction, this expression can be written as

$$
\begin{aligned}
\hat{u}_\xi &= \frac{E\delta_1^2}{6}\int_0^a \frac{(Q_\xi q-Q_\eta p)\,q-(Q_\xi q+Q_\eta p)\,q}{(Q_\xi q-Q_\eta p)^2}\,\mathrm{d}\alpha \\
&= \frac{E\delta_1^2}{6}\int_0^a \frac{-2Q_\eta pq}{(Q_\xi q-Q_\eta p)^2}\,\mathrm{d}\alpha
\end{aligned}
\tag{3.81}
$$

$$
\begin{aligned}
\hat{u}_\eta &= \frac{E\delta_1^2}{6} \int_0^a \frac{(Q_\xi q - Q_\eta p)\, p - (Q_\xi q + Q_\eta p)\,(-p)}{(Q_\xi q - Q_\eta p)^2}\, \mathrm{d}\alpha \\
&= \frac{E\delta_1^2}{6} \int_0^a \frac{2Q_\xi pq}{(Q_\xi q - Q_\eta p)^2}\, \mathrm{d}\alpha
\end{aligned}
\tag{3.82}
$$

The flexibility due to the crack is defined as $\hat{g}_i = \partial \hat{u}_i / \partial Q_i$

$$
\hat{g}_\xi = \frac{\partial \hat{u}_\xi}{\partial Q_\xi} = \frac{2E\delta_1^2}{3} \int_0^a \frac{Q_\eta pq^2}{(Q_\xi q - Q_\eta p)^3}\, \mathrm{d}\alpha
\tag{3.83}
$$

$$
\hat{g}_\eta = \frac{\partial \hat{u}_\eta}{\partial Q_\eta} = \frac{2E\delta_1^2}{3} \int_0^a \frac{Q_\eta p^2 q}{(Q_\xi q - Q_\eta p)^3}\, \mathrm{d}\alpha
\tag{3.84}
$$

Adding flexibility of the uncracked shaft to additional flexibility due to the crack, one obtains

$$
g_\xi = \frac{L^3}{48EI} + \frac{2E\delta_1^2}{3} \int_0^a \frac{Q_\eta pq^2}{(Q_\xi q - Q_\eta p)^3}\, \mathrm{d}\alpha
\tag{3.85}
$$

$$
g_\eta = \frac{L^3}{48EI} + \frac{2E\delta_1^2}{3} \int_0^a \frac{Q_\xi p^2 q}{(Q_\xi q - Q_\eta p)^3}\, \mathrm{d}\alpha
\tag{3.86}
$$

Using the above flexibility values, the stiffness values can be obtained directly.

As an illustrative example for crack depth $a = 0.1d$, adding flexibility to uncracked shaft to additional flexibility due to the crack, one obtains

$$
\hat{g}_\xi = \frac{2E\delta_1^2}{3m^2 g^2} \int_0^{2\pi} \frac{\sqrt{0.25d^2 - D_\theta^2}\, D_\theta^2 \sin\theta}{\left(-D_\theta \cos\theta - \sin\theta \sqrt{0.25d^2 - D_\theta^2}\right)^3} \left(-\frac{9d\sin\theta}{D_\theta}\, \mathrm{d}\theta\right)
\tag{3.87}
$$

$$
\hat{g}_\eta = \frac{2E\delta_1^2}{3m^2 g^2} \int_0^{2\pi} \frac{(0.25d^2 - D_\theta^2)\, D_\theta \cos\theta}{\left(-D_\theta \cos\theta - \sin\theta \sqrt{0.25d^2 - D_\theta^2}\right)^3} \left(-\frac{9d\sin\theta}{D_\theta}\, \mathrm{d}\theta\right)
\tag{3.88}
$$

where

$$
D_\theta = 0.05d\sqrt{82 + 18\cos\theta}
\tag{3.89}
$$

Finally, after numerical integration by using composite Simpson's rule in Eqs.(3.87) and (3.88), the stiffnesses of the cracked shaft are obtained as the inverse of the flexibility. Figure 3.18 shows the normalised stiffness variations for relative crack depth $a/d = 0.1$ in rotating coordinates. The normalised stiffness is compared with the model of Jun et al. [66]. For the proposed stiffness estimation based on CZM, the result is a nearly harmonic change of stiffness. However, in Jun's model, the stiffness change is not exactly harmonic, which is due to the fact that the partial opening and closing behaviour of the crack is linked to the sign of the SIFs.

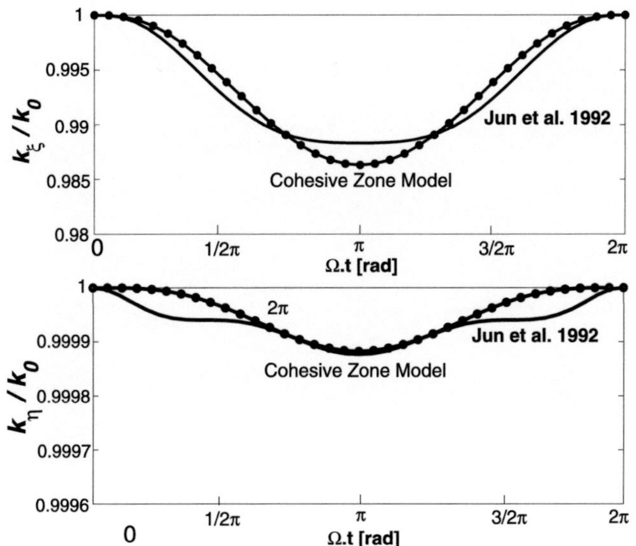

Figure 3.18: Normalised shaft stiffness variations of a cracked rotor with a breathing crack for relative crack depth $a/d = 0.1$

The curve fitting of the normalised shaft stiffnesses shown in Figure 3.19 yield the following relations

$$f_\xi(\Omega.t) = 0.9883 + 0.0123\cos(\Omega.t) \tag{3.90}$$

$$f_\eta(\Omega.t) = 0.9957 + 0.0051\cos(\Omega.t) \tag{3.91}$$

Figure 3.20 shows the normalised shaft stiffness with various relative crack depths a/d, which are calculated similarly. It is shown that shaft stiffness variation increases nonlinearly with crack depth.

The normalised shaft stiffness can be presented in the fixed coordinates as shown in Figure 3.21 using coordinates transformation based on Figure 3.1.

$$\left\{ \begin{array}{c} F_y \\ F_x \end{array} \right\} = \left[\begin{array}{cc} \cos\theta & -\sin\theta \\ \sin\theta & \cos\theta \end{array} \right] \left\{ \begin{array}{c} F_\xi \\ F_\eta \end{array} \right\} \tag{3.92}$$

$$\left[\begin{array}{cc} k_y & k_{yx} \\ k_{xy} & k_x \end{array} \right] \left\{ \begin{array}{c} y \\ z \end{array} \right\} = \left[\begin{array}{cc} \cos\theta & -\sin\theta \\ \sin\theta & \cos\theta \end{array} \right] \left[\begin{array}{cc} k_\xi & k_{\xi\eta} \\ k_{\eta\xi} & k_\eta \end{array} \right] \left\{ \begin{array}{c} \xi \\ \eta \end{array} \right\} \tag{3.93}$$

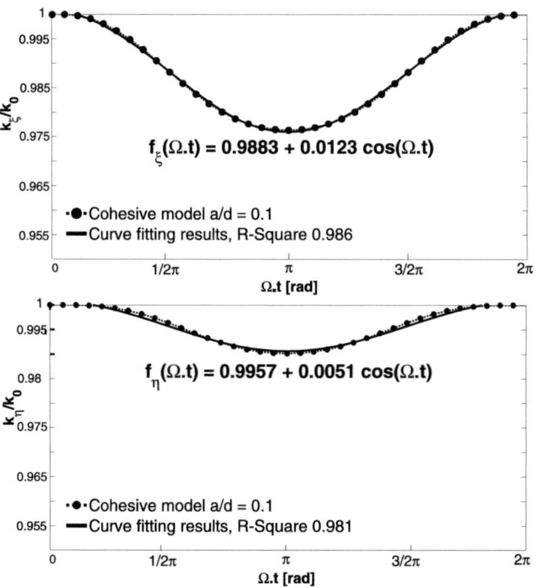

Figure 3.19: Curve fitting of normalised shaft stiffness for relative crack depth $a/d = 0.1$

Using Eq.(3.6), the stiffness coefficients in fixed coordinates can be written as

$$\begin{bmatrix} k_y & k_{yx} \\ k_{xy} & k_x \end{bmatrix} = \begin{bmatrix} \cos\theta & -\sin\theta \\ \sin\theta & \cos\theta \end{bmatrix} \begin{bmatrix} k_\xi & k_{\xi\eta} \\ k_{\eta\xi} & k_\eta \end{bmatrix} \begin{bmatrix} \cos\theta & \sin\theta \\ -\sin\theta & \cos\theta \end{bmatrix} \tag{3.94}$$

or

$$\begin{bmatrix} k_y & k_{yx} \\ k_{xy} & k_x \end{bmatrix} = \boldsymbol{T}^{-1} \begin{bmatrix} k_\xi & k_{\xi\eta} \\ k_{\eta\xi} & k_\eta \end{bmatrix} \boldsymbol{T} \tag{3.95}$$

where the transformation matrix is

$$\boldsymbol{T} = \begin{bmatrix} \cos\theta & \sin\theta \\ -\sin\theta & \cos\theta \end{bmatrix} \tag{3.96}$$

Substituting curve fitting results described by Eq.(3.90) and (3.91) into equations of motion of the system Eq.(3.34), the system is parametrically excited because the excitation appears in the coefficients of the governing differential equations. Figures 3.22 and 3.23 illustrate free vibration responses in rotating coordinates of Eq.(3.34) and Figure 3.24 represents steady state orbital responses compared to responses based on Mayes' model which the breathing function is described by cosine function (Eq.(3.19)).

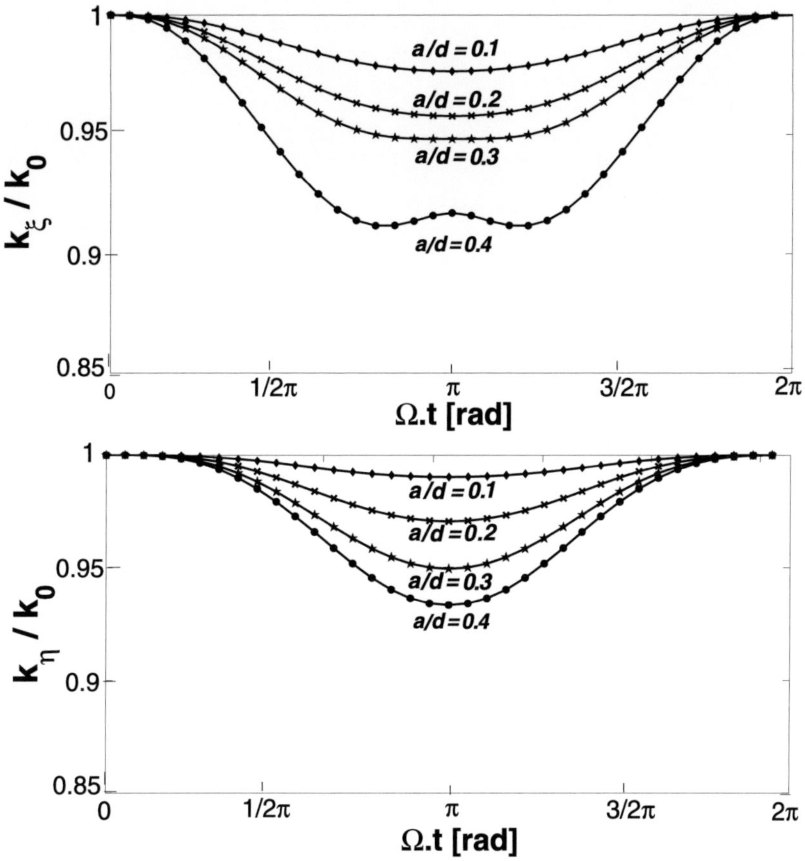

Figure 3.20: Normalised shaft stiffness vs. various relative crack depths a/d in rotating coordinates

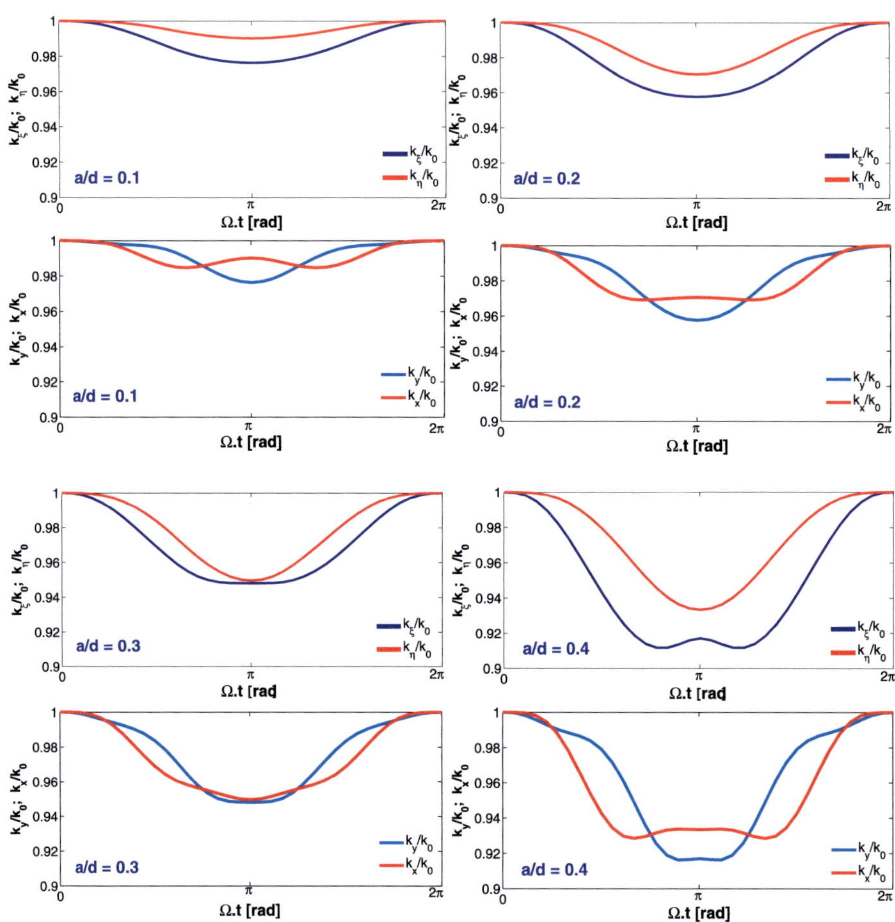

Figure 3.21: Normalised shaft stiffness vs. various relative crack depths a/d in fixed and rotating coordinates

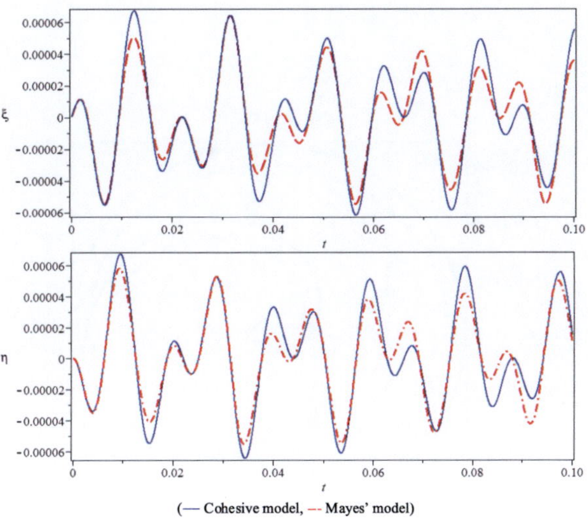

Figure 3.22: Vibration response in rotating coordinates for $a/d = 0.1$, $\Omega = 500$ rad/s

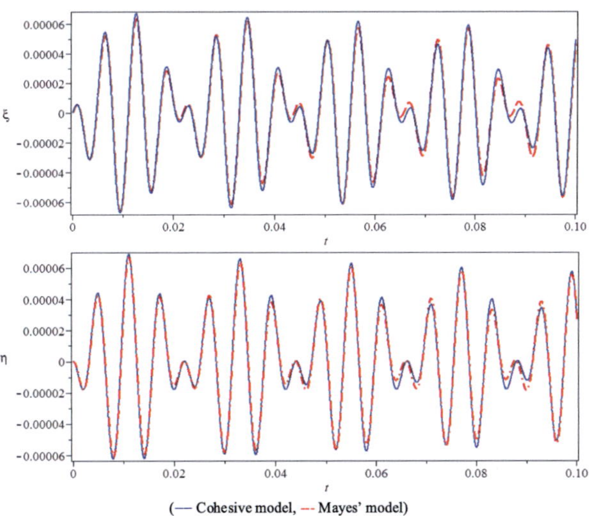

Figure 3.23: Vibration response in rotating coordinates for $a/d = 0.1$, $\Omega = 1000$ rad/s

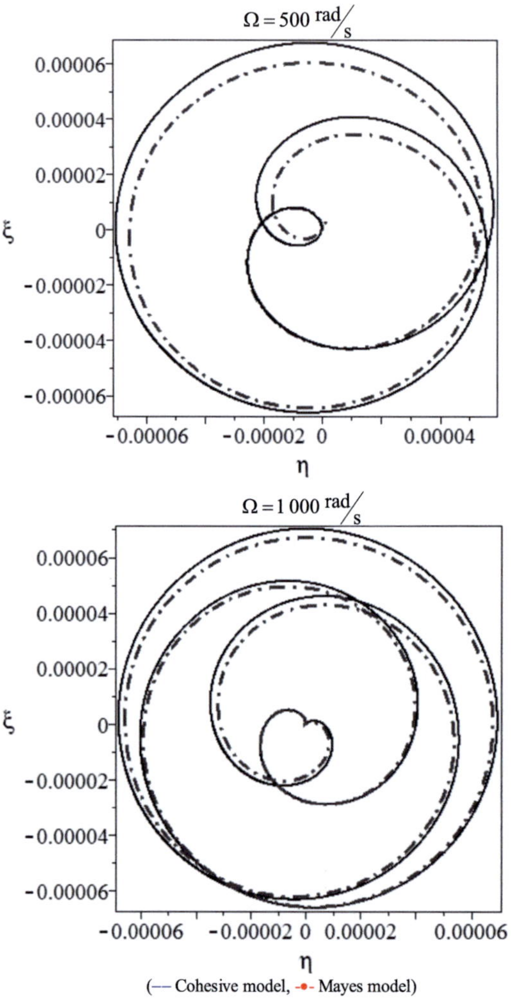

Figure 3.24: Steady state orbital responses in rotating coordinates for relative crack depth $a/d = 0.1$, $\Omega = 500$ rad/s and 1000 rad/s

3.5 Breathing crack with large crack depth ($a/d > 0.2$)

In order to obtain the stiffness variation for large crack depth ($a/d \geq 0.2$), the crack closure parabolic line model is applied [84]. The geometry of parabolic line is modelled as shown in Figure 3.25 where the parabolic line in half revolution of the shaft is divided in 9 lines, obtaining angular steps of $\pi/8$. This partition has been found to yield sufficiently accurate results. At the beginning ($n = 1$), crack is fully open, the parabolic line is a parallel line, and the crack width is maximum. At the end ($n = 9$), crack is fully closed and the parabolic line vanishes. For partially open crack $1 < n < 9$, the crack closure parabolic line propagates along ξ-axis.

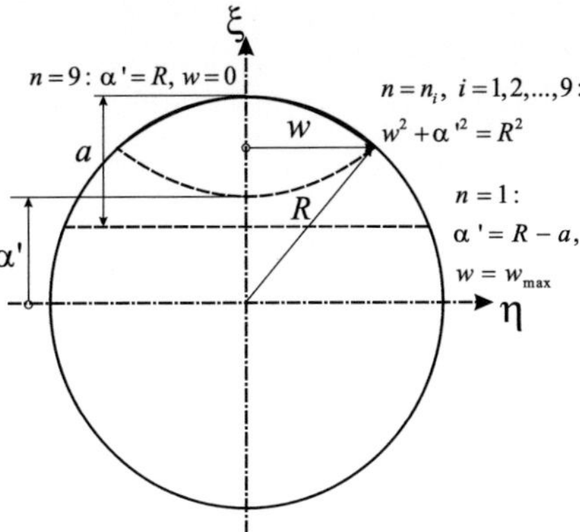

Figure 3.25: Geometry of crack closure parabolic line of the breathing cracked shaft

The breathing parabolic function is defined by

$$\xi_P = \left(R - \frac{9}{8}a\right) + \frac{1}{8}a \cdot n + \frac{1}{8}a \cdot (n-1)\left(\frac{\eta_P}{R}\right)^2 \qquad n = 1, 2, \ldots, 9 \tag{3.97}$$

where n is divided in 9 lines according angular steps of $\pi/8$ and radius of shaft $R = 0.5d$. Crack depth in the radial direction along ξ-axis can be written

$$\alpha' = \left(R - \frac{9}{8}a\right) + \frac{1}{8}a \cdot n + \frac{1}{8}a \cdot (n-1) \qquad n = 1, 2, \ldots, 9 \tag{3.98}$$

Crack width in the parallel direction along η-axis is determined by intersection between parabolic line Eq.(3.98) and equation of circular shaft

$$w^2 + \alpha'^2 = R^2 \tag{3.99}$$

$$w^2 + \left[\left(R - \frac{9}{8}a\right) + \frac{1}{8}a \cdot n + \frac{1}{8}a \cdot (n-1)\left(\frac{w}{R}\right)^2\right]^2 = R^2 \tag{3.100}$$

Table 3.1: Crack states for half angle rotation $\Omega t = 0 \div \pi$ for $a/d = 0.2$

n	Ωt	State of crack $\%A_{cr} = A_{cr}/A$	Crack depth α'	Crack width w $w^2 + \alpha'^2 = R^2$
1	π	Fully open : 5.20%A	0.800 R	0,6 R
2	$7\pi/8$	Partially open : 4.28%A	0.825 R	0,533 774 R
3	$3\pi/4$	Partially open : 3.40%A	0.850 R	0,505 579 R
4	$5\pi/8$	Partially open : 2.59%A	0.875 R	0,463 980 R
5	$\pi/2$	Partially open : 1.85%A	0.900 R	0,400 996 R
6	$3\pi/8$	Partially open : 1.20%A	0.925 R	0,342 177 R
7	$\pi/4$	Partially open : 0.65%A	0.950 R	0,275 271 R
8	$\pi/8$	Partially open : 0.23%A	0.975 R	0,191 786 R
9	0	Fully closed : 0%A	R	0

Table 3.1, gives example for some states of crack during half rotation of shaft using the crack closure parabolic line for relative crack depth $a/d = 0.2$.

As the shaft rotates, the crack takes on different angular positions at the cracked section. From the transverse forces Q_ξ and Q_η acting on the same strip, the nominal stresses are given by

$$\sigma_\xi = \frac{16Q_\xi L}{\pi d^4}\alpha' \tag{3.101}$$

$$\sigma_\eta = \frac{16Q_\eta L}{\pi d^4}w \tag{3.102}$$

In the bent shaft while rotating, the forces Q_ξ and Q_η acting along the ξ and η axes on the cross-section containing the crack induce deflections of the solid shaft. The additional deflections due to the crack are estimated using the CZM. Similarly procedure as in crack closure straight line model (Section 3.4) is applied to estimate the stiffness variation of the cracked shaft.

The stress ratio can be written as

$$r_\sigma = \frac{\sigma_1}{\sigma_2} = \frac{\sigma_\eta}{\sigma_\xi} = \frac{Q_\eta w}{Q_\xi \alpha'} \tag{3.103}$$

The crack is fully closed at $\alpha' = R$ and $w = 0$, and the crack is fully open at $\alpha' = R - a$ and $w = w_{max}$. Therefore, the additional deflection is given by

$$\hat{u}_i = \frac{\partial}{\partial Q_i}\left[\int_0^a \frac{E\delta_1^2}{3}\left(1 + \frac{1}{2}\frac{Q_\xi \alpha' + Q_\eta w}{Q_\xi \alpha' - Q_\eta w} - \frac{1}{\sqrt{3}}\ln\left(\frac{S\sigma_Y}{CE}\right)\right)\,\mathrm{d}\alpha\right] \tag{3.104}$$

In ξ and η direction, this expression can be written as Eqs.(3.85-86), then the flexibility due to the crack is defined as $\hat{g}_i = \partial \hat{u}_i / \partial Q_i$ and addding flexibility of the uncracked shaft to the additional flexibility due to crack yields

$$g_\xi = \frac{L^3}{48EI} + \frac{2E\delta_1^2}{3} \int_{R-a}^{R} \frac{Q_\eta w \alpha'^2}{(Q_\xi \alpha' - Q_\eta w)^3} \, d\alpha \tag{3.105}$$

$$g_\eta = \frac{L^3}{48EI} + \frac{2E\delta_1^2}{3} \int_{R-a}^{R} \frac{Q_\eta w \alpha'^2}{(Q_\xi \alpha' - Q_\eta w)^3} \, d\alpha \tag{3.106}$$

Using the above flexibility values, the shaft stiffness can be obtained in each direction of ξ and η axes. After numerical integration, the stiffness of the cracked shaft (the inverse of the flexibility) can be calculated. Figure 3.26 represents the normalised stiffness variations for relative crack depth a/d from 0.1 to 0.4 in rotating coordinates using crack closure parabolic line model. The normalised shaft stiffnesses based on crack closure parabolic line model are compared with the normalised shaft stiffnesses based on straight line crack closure line model discussed in Section 3.4. For the crack closure straight line model, the result is a nearly harmonic change of stiffness. However, for the crack closure parabolic line model, the varying stiffness is not exactly harmonic, except for shallow cracks.

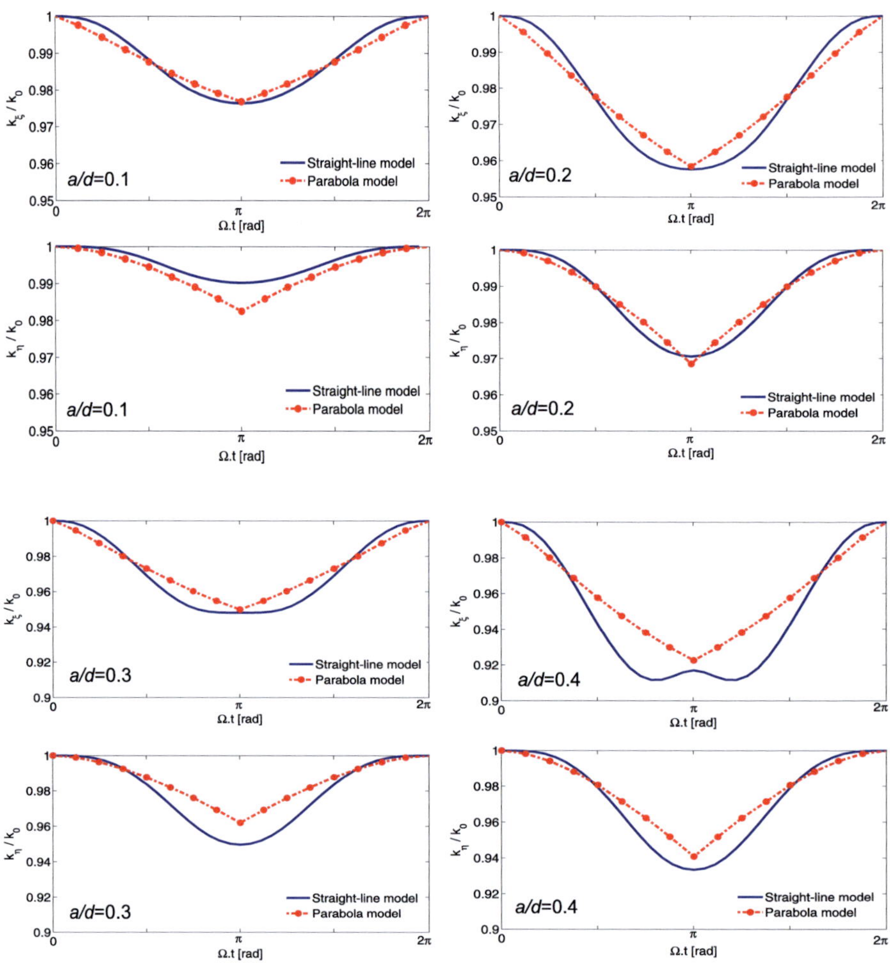

Figure 3.26: Normalised shaft stiffnesses for relative crack depth $a/d = 0.1 \div 0.4$

4 Stability analysis of a rotor with a transverse breathing crack

Presence of cracks in rotors leads to instability problems due to the local aymmetry introduced by the breathing mechanism of a crack. This chapter deals with the stability of simple rotor system (de Laval rotor) due to a breathing crack. To focus on crack influence alone, crack-disk imbalance interaction and internal damping are ignored and the perturbation method is used to obtain the boundaries of stability regions. It can be observed that some small damping in the rotor system is very helpful to guarantee stability.

4.1 Stability of rotor systems

All definitions of stability are concerned with the response of a system to certain disturbances and whether or not the response stays within certain bounds. If a linear, stationary rotor has symmetric, positive definite mass, stiffness and damping matrices, it will be stable and all external perturbations will decay back to the reference state. A shaft becomes asymmetric when a crack appears and grows, and it is of interest to know the consequences. For simple rotor systems there are three basic reasons for instability: asymmetric shaft stiffness, internal damping and cross-coupling stiffness and damping of bearings.

- Analysis shows that the symmetric component of the stiffness matrix produces conservative restoring forces and its element are properly called stiffnesses. The skew-symmetric component produces non-conservative tangential forces that can insert power into the rotor during each revolution and hence it can be destabilizing [96].

- Internal damping is due primarily to friction at rotor component interfaces. When the power inserted by internal damping exceeds the power extracted by external damping (support bearing damping), the rotor will become unstable.

- Fluid stiffness and damping cross coupling occurs in many bearings and some seals. The force associated with cross coupling stiffness can insert power into the rotor and if this power exceeds the power extracted by the direct fluid damping forces, the rotor will become unstable. Bearing clearance plays an important role in the stability of rotors. Increasing bearing clearance reduces the cross-coupling stiffnesses but it also reduces the external damping, making internal damping more of an instability threat.

To focus the crack influence alone on stability, rigid bearings are used and the internal damping interaction is ignored. When the symmetric shaft on rigid bearings is considered, it is easier to write the equations in rotating coordinates. However, when the bearing forces are to be combined to the asymmetric shaft, it becomes useful to write the equations in fixed coordinate system [113].

4.2 Instability due to parametric excitation

In order to analyse the effects of a crack in the shaft, the equations of motion should be formulated with consideration of the stiffness modification. Here, the direct stiffnesses k_ξ and k_η, are considered only. The cross-coupled stiffnesses remain relatively small compared with the direct stiffness [66], [76]. In this section, the boundaries of the stability regions of periodic solutions of linear rotor-system equations are obtained. The equations of motion for the rotor with a breathing crack which are discussed in previous chapter (Eqs.(3.36-37)) can be written

$$m\left(\ddot{\xi} - 2\dot{\theta}\dot{\eta} - \dot{\theta}^2\xi\right) + c\left(\dot{\xi} - \dot{\theta}\eta\right) + k_\xi\xi = 0 \tag{4.1}$$

$$m\left(\ddot{\eta} + 2\dot{\theta}\dot{\xi} - \dot{\theta}^2\eta\right) + c\left(\dot{\eta} + \dot{\theta}\xi\right) + k_\eta\eta = 0 \tag{4.2}$$

For simplicity, it is assumed that damping is neglected and for synchronous whirl $\dot{\theta} = \Omega$, hence

$$\ddot{\xi} - 2\Omega\dot{\eta} - \Omega^2\xi + \frac{k_\xi}{m}\xi = 0 \tag{4.3}$$

$$\ddot{\eta} + 2\Omega\dot{\xi} - \Omega^2\eta + \frac{k_\eta}{m}\eta = 0 \tag{4.4}$$

From Eqs.(3.90-91) for relative crack depth $a/d = 0.1$

$$\frac{k_\xi}{m} = (a_\xi + b_\xi \cos \Omega t)\frac{k}{m} = (0.988 + 0.012 \cos \Omega t)\frac{k}{m} \tag{4.5}$$

$$\frac{k_\eta}{m} = (a_\eta + b_\eta \cos \Omega t)\frac{k}{m} = (0.995 + 0.005 \cos \Omega t)\frac{k}{m} \tag{4.6}$$

Therefore

$$\ddot{\xi} + \left[\left(a_\xi\frac{k}{m} - \Omega^2\right) + b_\xi\frac{k}{m}\cos \Omega t\right]\xi - 2\Omega\dot{\eta} = 0 \tag{4.7}$$

$$\ddot{\eta} + \left[\left(a_\eta\frac{k}{m} - \Omega^2\right) + b_\eta\frac{k}{m}\cos \Omega t\right]\eta + 2\Omega\dot{\xi} = 0 \tag{4.8}$$

By introducing

$$\omega_{0\xi}^2 = a_\xi \frac{k}{m} - \Omega^2 \tag{4.9}$$

$$\omega_{0\xi}^2 \varphi_\xi = b_\xi \frac{k}{m} \tag{4.10}$$

$$\omega_{0\eta}^2 = a_\eta \frac{k}{m} - \Omega^2 \tag{4.11}$$

$$\omega_{0\eta}^2 \varphi_\eta = b_\eta \frac{k}{m} \tag{4.12}$$

Finally, the linear-coupled Mathieu-equations can be written

$$\ddot{\xi} + \omega_{0\xi}^2 \left(1 + \varphi_\xi \cos \Omega t\right) \xi - 2\Omega \dot{\eta} = 0 \tag{4.13}$$

$$\ddot{\eta} + \omega_{0\eta}^2 \left(1 + \varphi_\eta \cos \Omega t\right) \xi + 2\Omega \dot{\xi} = 0 \tag{4.14}$$

The stability problem for the linear coupled Mathieu-equations has been studied by many authors (Hansen [55], Mahmoud [87], Takahashi [139], Ikeda and Murakami [60] and Wettergren and Olson [150]). It is known that for the instability domains corresponding to the natural frequencies, the boundary curves can be found by searching for the periodic solutions of Eqs.(4.13-14). In the literature [52], [93], [151] the stability of the Mathieu-equations are investigated directly by using Floquet theory and any one of the following techniques: perturbations, Fourier analysis or numerical integration. In this work, the boundaries of stability regions for periodic solutions of linear-coupled Mathieu equations are obtained by second-order perturbation method.

Consider now the case that φ small (Eqs.(4.13-14)) in which the perturbation method is applicable. By substituting

$$\Omega t = 2\tau \tag{4.15}$$

where

$$\dot{\xi} = \frac{d\xi}{dt} = \frac{d\xi}{d\tau}\frac{d\tau}{dt} = \frac{1}{2}\Omega\xi' \tag{4.16}$$

$$\ddot{\xi} = \frac{d^2\xi}{dt^2} = \frac{d}{dt}\left(\frac{d\xi}{dt}\right) = \frac{d}{dt}\left(\frac{1}{2}\Omega\xi'\right) = \frac{1}{4}\Omega^2\xi'' \tag{4.17}$$

and similar with $\ddot{\eta}$ and $\dot{\eta}$ with $(\cdot)$ as $d\,(\,)/dt$ and $(')$ as $d\,(\,)/d\tau$, then

$$\xi'' + \frac{4\omega_{0\xi}^2}{\Omega^2}\left(1 + \varphi_\xi \cos \Omega t\right) \xi - 4\eta' = 0 \tag{4.18}$$

$$\eta'' + \frac{4\omega_{0\eta}^2}{\Omega^2}\left(1 + \varphi_\eta \cos \Omega t\right) \xi + 4\xi' = 0 \tag{4.19}$$

Define

$$\delta_\xi = \left(\frac{2\omega_{0\xi}}{\Omega}\right)^2 \tag{4.20}$$

$$\delta_\eta = \left(\frac{2\omega_{0\eta}}{\Omega}\right)^2 \tag{4.21}$$

$$\varepsilon_\xi = \left(\frac{2\omega_{0\xi}}{\Omega}\right)^2 \varphi_\xi \tag{4.22}$$

$$\varepsilon_\eta = \left(\frac{2\omega_{0\eta}}{\Omega}\right)^2 \varphi_\eta \tag{4.23}$$

Finally, the linear-coupled Mathieu equations can be written in the form

$$\xi'' + (\delta_\xi + \varepsilon_\xi \cos 2\tau)\, \xi - 4\eta' = 0 \tag{4.24}$$

$$\eta'' + (\delta_\eta + \varepsilon_\eta \cos 2\tau)\, \eta + 4\xi' = 0 \tag{4.25}$$

These equations are investigated directly by considering solutions of Eqs.(4.24-25) which oscillate with the unperturbed frequencies [52], for the solution

$$\delta_\xi = n^2 + \varepsilon_\xi \delta_{\xi 1} + \varepsilon_\xi^2 \delta_{\xi 2} + \cdots \tag{4.26}$$

$$\delta_\eta = m^2 + \varepsilon_\eta \delta_{\eta 1} + \varepsilon_\eta^2 \delta_{\eta 2} + \cdots \tag{4.27}$$

and the solution can be written as a series expansion in ε

$$\xi(\tau) = \xi_0(\tau) + \varepsilon_\xi \xi_1(\tau) + \varepsilon_\xi^2 \xi_2(\tau) + \cdots \tag{4.28}$$

$$\eta(\tau) = \eta_0(\tau) + \varepsilon_\eta \eta_1(\tau) + \varepsilon_\eta^2 \eta_2(\tau) + \cdots \tag{4.29}$$

As an example the case $n = m$ is investigated, as other values of n and m can be similarly treated. The approximate stability regions are valid only for small value of parameter ε ($\varepsilon < 0.2$). Suppose that the damping coefficient is of order ε^2, i.e. $h = 4$ in Eqs.(4.24-25) is of order ε^2, then we can write $h = \varepsilon_\xi^2 h_1 = \varepsilon_\eta^2 h_2$, where h_1 and h_2 have maximal value about 100. Substituting from Eqs.(4.28-29) into Eqs.(4.24-25), yield

$$\xi_0'' + \varepsilon_\xi \xi_1'' + \varepsilon_\xi^2 \xi_2'' + \cdots + \left(n^2 + \varepsilon_\xi \delta_{\xi 1} + \varepsilon_\xi^2 \delta_{\xi 2} + \cdots + \varepsilon_\xi \cos 2\tau\right)$$
$$\left(\xi_0 + \varepsilon_\xi \xi_1 + \varepsilon_\xi^2 \xi_2 + \cdots\right) - \varepsilon_\xi^2 h_1 \eta_0' = 0 \tag{4.30}$$

$$\eta_0'' + \varepsilon_\eta \eta_1'' + \varepsilon_\eta^2 \eta_2'' + \cdots + \left(m^2 + \varepsilon_\eta \delta_{\eta 1} + \varepsilon_\eta^2 \delta_{\eta 2} + \cdots + \varepsilon_\eta \cos 2\tau\right)$$
$$\left(\eta_0 + \varepsilon_\eta \eta_1 + \varepsilon_\eta^2 \eta_2 + \cdots\right) - \varepsilon_\eta^2 h_2 \xi_0' = 0 \tag{4.31}$$

Equating like powers of ε yields a doubly infinite hierarchy of equations at order ε^j, for ξ_j and η_j, $j{=}0,1,2,\ldots$, as follows

$$\xi_0'' + n^2\xi_0 = 0 \tag{4.32}$$

$$\xi_1'' + n^2\xi_1 = -\delta_{\xi 1}\xi_0 - \xi_0\cos 2\tau \tag{4.33}$$

$$\xi_2'' + n^2\xi_2 = -\delta_{\xi 1}\xi_1 - \delta_{\xi 2}\xi_0 - \xi_1\cos 2\tau + h_1\eta_0' \tag{4.34}$$

$$\eta_0'' + m^2\eta_0 = 0 \tag{4.35}$$

$$\eta_1'' + m^2\eta_1 = -\delta_{\eta 1}\eta_0 - \eta_0\cos 2\tau \tag{4.36}$$

$$\eta_2'' + m^2\eta_2 = -\delta_{\eta 1}\eta_1 - \delta_{\eta 2}\eta_0 - \eta_1\cos 2\tau - h_2\xi_0' \tag{4.37}$$

Case $n = 1$

Solution to Eq.(4.32) and Eq.(4.35) can be written as

$$\xi_0 = a\cos\tau + b\sin\tau \tag{4.38}$$

$$\eta_0 = a\sin\tau - b\cos\tau \tag{4.39}$$

Inserting Eq.(4.38) into Eq.(4.33), we have

$$\begin{aligned}
\xi_1'' + \xi_1 &= -\delta_{\xi 1}\left(a\cos\tau + b\sin\tau\right) - \left(a\cos\tau + b\sin\tau\right)\cos 2\tau \\
&= \left(-\delta_{\xi 1} - \frac{1}{2}\right)a\cos\tau + \left(-\delta_{\xi 1} + \frac{1}{2}\right)b\sin\tau \\
&\quad - \frac{1}{2}a\cos 3\tau - \frac{1}{2}b\cos 3\tau
\end{aligned} \tag{4.40}$$

In order to eliminate secular terms, we must have

$$-\delta_{\xi 1} - \frac{1}{2} = 0 \tag{4.41}$$

or

$$-\delta_{\xi 1} + \frac{1}{2} = 0 \tag{4.42}$$

As a result, in order to avoid ξ_0 being the zero solution, we must have $\delta_{\xi 1} = \pm\frac{1}{2}$, then

$$\xi_1'' + \xi_1 = -\frac{1}{2}a\cos 3\tau - \frac{1}{2}b\sin 3\tau \tag{4.43}$$

and the solution is

$$\xi_1 = \frac{1}{16}\left(a\cos 3\tau + b\sin 3\tau\right) \tag{4.44}$$

Finally, inserting Eqs.(4.44) and (4.39) into Eq.(4.34), we obtain

$$
\begin{aligned}
\xi_2'' + \xi_2 &= \frac{1}{32}\left(a\cos 3\tau + b\sin 3\tau\right) - \delta_{\xi 2}\left(a\cos\tau + b\sin\tau\right) - \\
&\quad \frac{1}{16}\left(a\cos 3\tau + b\sin 3\tau\right)\cos 2\tau + h_1\left(a\cos\tau + b\sin\tau\right) \\
&= \left(-\delta_{\xi 2} - \frac{1}{32} + h_1\right)a\cos\tau + \left(-\delta_{\xi 2} - \frac{1}{32} + h_1\right)b\sin\tau + \\
&\quad \frac{1}{32}a\left(\cos 3\tau - \cos 5\tau\right) + \frac{1}{32}b\left(\sin 3\tau - \sin 5\tau\right)
\end{aligned}
\tag{4.45}
$$

As a result, in order to avoid a nonzero ξ_0, we must have

$$
\delta_{\xi 2} = -\frac{1}{32} + h_1
\tag{4.46}
$$

Then, we get

$$
\xi_2'' + \xi_2 = \frac{1}{32}a\left(\cos 3\tau - \cos 5\tau\right) + \frac{1}{32}b\left(\sin 3\tau - \sin 5\tau\right)
\tag{4.47}
$$

and the solution is

$$
\xi_2 = \frac{1}{256}\left(a\cos 3\tau + b\sin 3\tau\right) - \frac{1}{768}\left(a\cos 5\tau + b\sin 5\tau\right)
\tag{4.48}
$$

Therefore

$$
\begin{aligned}
\delta_\xi &= 1 \pm \frac{1}{2}\varepsilon_\xi + \left(-\frac{1}{32} + h_1\right)\varepsilon_\xi^2 \\
&= 5 \pm \frac{1}{2}\varepsilon_\xi - \frac{1}{32}\varepsilon_\xi^2 + \mathcal{O}\left(\varepsilon_\xi^3\right)
\end{aligned}
\tag{4.49}
$$

We also have either

$$
\begin{aligned}
\xi(\tau) &= \xi_0(\tau) + \varepsilon_\xi\xi_1(\tau) + \varepsilon_\xi^2\xi_2(\tau) + \cdots \\
&= a\cos\tau + \frac{1}{16}\varepsilon_\xi a\cos 3\tau + \frac{1}{768}\varepsilon_\xi^2 a\left(3\cos 3\tau - \cos 5\tau\right) + \mathcal{O}\left(\varepsilon_\xi^3\right)
\end{aligned}
\tag{4.50}
$$

or

$$
\xi(\tau) = b\sin\tau + \frac{1}{16}\varepsilon_\xi b\sin 3\tau + \frac{1}{768}\varepsilon_\xi^2 b\left(3\sin 3\tau - \sin 5\tau\right) + \mathcal{O}\left(\varepsilon_\xi^3\right)
\tag{4.51}
$$

Case $n = 2$

Solution to Eq.(4.32) and Eq.(4.35) can be written as

$$\xi_0 = a\cos 2\tau + b\sin 2\tau \tag{4.52}$$

$$\eta_0 = a\sin 2\tau - b\cos 2\tau \tag{4.53}$$

Inserting Eq.(4.52) into Eq.(4.33), we have

$$\begin{aligned}
\xi_1'' + 4\xi_1 &= -\delta_{\xi 1}\left(a\cos 2\tau + b\sin 2\tau\right) - \left(a\cos 2\tau + b\sin 2\tau\right)\cos 2\tau \\
&= -\delta_{\xi 1}a\cos 2\tau - \delta_{\xi 1}b\sin 2\tau - \frac{1}{2}a - \frac{1}{2}a\cos 4\tau - \frac{1}{2}b\cos 4\tau
\end{aligned} \tag{4.54}$$

In order to eliminate secular terms, we must have

$$\delta_{\xi 1} = 0 \tag{4.55}$$

As a result

$$\xi_1'' + 4\xi_1 = -\frac{1}{2}a - \frac{1}{2}a\cos 4\tau - \frac{1}{2}b\sin 4\tau \tag{4.56}$$

and the solution is

$$\xi_1 = -\frac{1}{8}a + \frac{1}{24}\left(a\cos 4\tau + b\sin 4\tau\right) \tag{4.57}$$

Inserting Eqs.(4.57) and (4.53) into Eq.(4.34), we have

$$\begin{aligned}
\xi_2'' + 4\xi_2 &= \delta_{\xi 1}\left(-\frac{1}{8}a + \frac{1}{24}a\cos 4\tau + \frac{1}{24}b\sin 4\tau\right) - \delta_{\xi 2}\left(a\cos 2\tau + b\sin 2\tau\right) - \\
&\quad \left(-\frac{1}{8}a + \frac{1}{24}a\cos 4\tau + \frac{1}{24}b\sin 4\tau\right)\cos 2\tau + h_1\left(2a\cos 2\tau + 2b\sin 2\tau\right) \\
&= \left(-\delta_{\xi 2} + \frac{5}{48} + 2h_1\right)a\cos 2\tau + \left(-\delta_{\xi 2} - \frac{1}{48} + 2h_1\right)b\sin 2\tau - \\
&\quad \frac{1}{48}a\left(\cos 6\tau + b\sin 6\tau\right)
\end{aligned} \tag{4.58}$$

As a result, in order to avoid a nonzero ξ_0, we must have

$$\delta_{\xi 2} = \frac{5}{48} + 2h_1 \tag{4.59}$$

or

$$\delta_{\xi 2} = -\frac{1}{48} + 2h_1 \tag{4.60}$$

Then, we obtain

$$\xi_2'' + 4\xi_2 = -\frac{1}{48}\left(a\cos 6\tau + b\cos 6\tau\right) \tag{4.61}$$

and the solution is

$$\xi_2 = \frac{1}{1536}\left(a\cos 6\tau + b\sin 6\tau\right) \tag{4.62}$$

Therefore

$$\delta_{\xi 1} = 2^2 - \frac{1}{48}\varepsilon_\xi^2 + 8 + \mathcal{O}\left(\varepsilon_\xi^3\right) \tag{4.63}$$

$$\delta_{\xi 2} = 2^2 + \frac{5}{48}\varepsilon_\xi^2 + 8 + \mathcal{O}\left(\varepsilon_\xi^3\right) \tag{4.64}$$

We also have either

$$
\begin{aligned}
\xi(\tau) &= \xi_0(\tau) + \varepsilon_\xi\xi_1(\tau) + \varepsilon_\xi^2\xi_2(\tau) + \cdots \\
&= a\cos 2\tau + \frac{1}{24}\varepsilon_\xi a\left(-3 + \cos 4\tau\right) + \frac{1}{1536}\varepsilon_\xi^2 a\cos 6\tau + \mathcal{O}\left(\varepsilon_\xi^3\right)
\end{aligned}
\tag{4.65}
$$

or

$$\xi(\tau) = b\sin 2\tau + \frac{1}{24}\varepsilon_\xi b\sin 4\tau + \frac{1}{1536}\varepsilon_\xi^2 b\sin 6\tau + \mathcal{O}\left(\varepsilon_\xi^3\right) \tag{4.66}$$

Case $n = 0$

Solution to Eq.(4.32) and Eq.(4.35) can be written as

$$\xi_0 = a + b\tau \tag{4.67}$$

$$\eta_0 = a + b\left(\tau - \frac{\pi}{2}\right) \tag{4.68}$$

Inserting Eq.(4.67) into Eq.(4.33), we have

$$\xi_1'' = -\delta_{\xi 1}a - a\cos 2\tau \tag{4.69}$$

In order to eliminate secular terms, we must have

$$
\begin{aligned}
\delta_{\xi 1}a &= 0 \\
\Leftrightarrow \delta_{\xi 1} &= 0
\end{aligned}
\tag{4.70}
$$

As a result

$$\xi_1'' = -a\cos 2\tau \tag{4.71}$$

$$\xi_1 = \frac{1}{4}a\cos 2\tau \tag{4.72}$$

Inserting Eqs.(4.72) and (4.68) into Eq.(4.34) yields

$$\xi_2'' = -\delta_{\xi 1}(0) - \delta_{\xi 2}a - \left(-\frac{1}{4}a\cos 2\tau\right)\cos 2\tau + h_1(0)$$

$$= -\delta_{\xi 2}a - \frac{1}{8}a - \frac{1}{8}a\cos 4\tau \tag{4.73}$$

As a result, in order to avoid a nonzero ξ_0, we must have

$$-\delta_{\xi 2}a - \frac{1}{8}a = 0 \tag{4.74}$$

$$\Leftrightarrow \delta_{\xi 2} = -\frac{1}{8} \tag{4.75}$$

Then, we obtain

$$\xi_2'' = -\frac{1}{8}a\cos 4\tau \tag{4.76}$$

and the solution is

$$\xi_2 = \frac{1}{128}a\cos 4\tau \tag{4.77}$$

Therefore

$$\delta_\xi = -\frac{1}{8}\varepsilon_\xi^2 + \mathcal{O}\left(\varepsilon_\xi^3\right) \tag{4.78}$$

We also have either

$$\xi(\tau) = \xi_0(\tau) + \varepsilon_\xi\xi_1(\tau) + \varepsilon_\xi^2\xi_2(\tau) + \cdots$$

$$= a + \frac{1}{4}\varepsilon_\xi a\cos 2\tau + \frac{1}{128}\varepsilon_\xi^2 a\cos 4\tau + \mathcal{O}\left(\varepsilon_\xi^3\right) \tag{4.79}$$

Expression of Eq.(4.49), Eqs.(4.63-64) and Eq.(4.78) are plotted as solid curve in Figure 4.1. The approximate stability regions are valid only for small value of parameter ε ($\varepsilon < 0.2$). Since practically the values ω_ξ and ω_η are very close to each other and since the rotational speeds $\Omega = \omega_\xi$ and $\Omega = \omega_\eta$ must be avoided as operational speeds.

According Eqs.(4.49), (4.63-64), (4.78), (4.20) and (4.9) for $\varepsilon \cong 0$, we have

$$n = 0, \qquad \Omega \approx \omega_n \tag{4.80}$$

$$n = 1, \qquad \Omega \approx 0.67\omega_n \tag{4.81}$$

$$n = 0, \qquad \Omega \approx 0.50\omega_n \tag{4.82}$$

Figure 4.2 depicts the instability regions at $\Omega/\omega_n = \frac{1}{2}, \frac{2}{3}, 1$ are very small. As soon as we have some damping, they will disappear. Some small damping in the system is very helpful to guarantee stability, concerning this compare the width of the unstable zones for damping and undamping regions in Figure 4.2. In real rotating machines, damping is normally sufficient to avoid the instability when a crack is present [12].

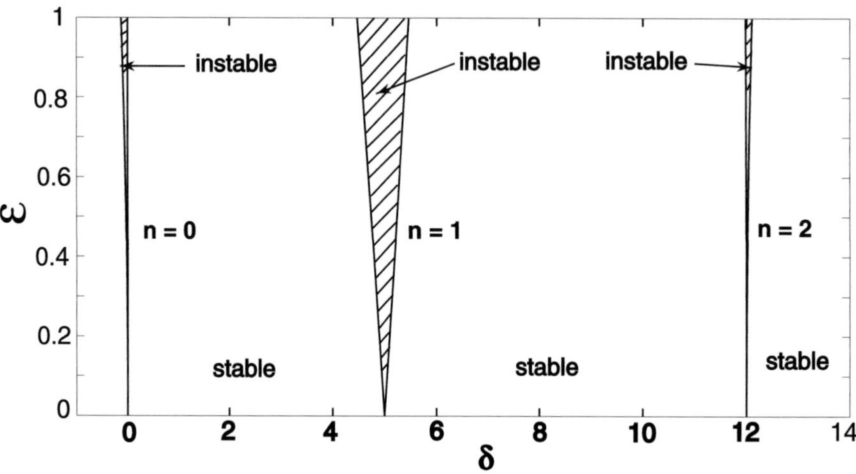

Figure 4.1: Approximations to the boundary between stability and instability of the linear-coupled Mathieu equations for $n=0, 1$ and 2

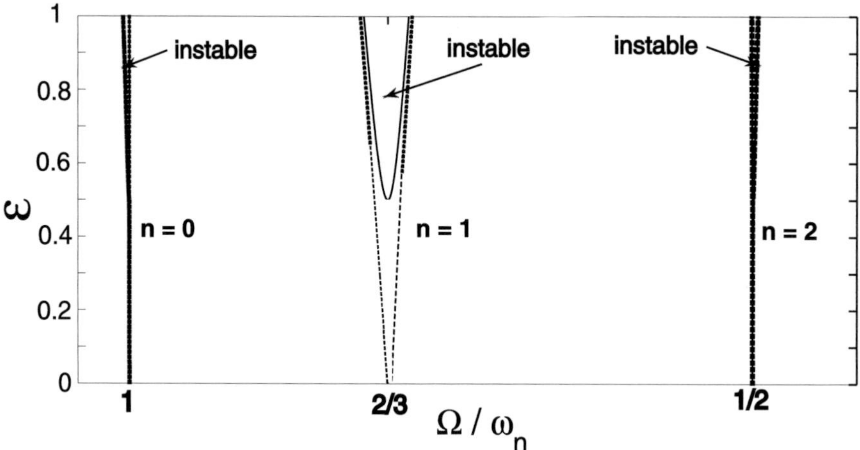

Figure 4.2: Borderlines of stability with a damping of 5%

5 Finite element model of the cracked shaft

This chapter presents the finite element (FE) modelling of a breathing cracked rotor, using an approach that has been developed by considering one dimensional continuum of shaft (equivalent beam), in which Timoshenko beam approach is used and the different second moment of area of the cross section is introduced. This approach is obviously approximated, but comparison with 3D calculation results show very good agreement, as long as single cracks with regular shapes (rectilinear or elliptical) are considered [11]. Also, the approach based on a zero thickness cohesive element as a model of fracture process zone.

In the first section, the cracked shaft without disk has been investigated. This condition is called "weight governed breathing" where the stresses due to dynamic loads are small with respect to those due to static loads, and therefore it can be neglected. Then, the breathing becomes a function of the angular position $\theta = \Omega t$ of the shaft only. Therefore the stiffness variation becomes periodical and independent of rotating speed and of exciting forces. Only horizontal and vertical displacement and angular deflection around horizontal and vertical axes are considered. The expressions for a constant section shaft element are considered taking into account transversal and rotational inertia and shear deformation. The elastic shaft is modelled including gyroscopic effects and the bearings are assumed as rigid supports in order to focus on crack influence.

The aims of this chapter is to model the cracked shaft based on CZM in conjunction with the FE. Two FE models using CZM are proposed, the first model is based on different asymmetric area moments of inertia of the cross section due to the crack, which depend on the breathing function in terms of the angular position obtained by curve fitting, as discussed in Chapter 3. The other model is based on the TSL using one element having zero thickness which is placed between the continuum elements. Results obtained from CZM are compared with those obtained from the proposed cosine function model by Mayes and Davis [89]. The analytical results based on the Timoshenko open cracked shaft are also reported. The last section of this chapter deals with FE modelling of the cracked rotor supported by rigid bearing with disk. It can be seen that the synchronous response has two contributions, one caused by the crack and one by the unbalance. The resultant of this superposition depends on the angular position between eccentricity and crack. The analytical result based on Dunkerley's equation and Rayleigh's method for open cracked shaft are also described. These methods are convenient to estimate the lower and upper bound of the natural frequencies and can be used as a reference in order to validate the proposed FE results.

5.1 Model of the rotor supported by rigid bearings

In most models used to study the vibrational behaviour of rotors, the actual continuous media system is modelled by a discrete assemblage. Thus, the governing partial differential equation embodying the applicable physical principles of the continuous structure is approximated by a set of ordinary differential equations. The fundamental reason is due to the fact that most governing partial differential equations can be derived only for the simplest geometric of shapes and are not suited to account for the complex shapes and configurations either for shafts-rotors or supports.

The system to be analysed consists of three main parts: rotor, disk and bearings. The rotor shaft is considered to be a flexible body with distributed mass and elasticity. The disk is assumed to be thin and rigid. Bearings are assumed to be rigid supports (very high value of stiffness) in order to focus on crack influence and to be able to compare with some published results. In order to develop the governing equations for the cracked rotor system, the equations for these components are first derived, and then the total system of equations is obtained by assembling the equations of the individual components.

5.1.1 Shaft model

The rotor is composed of a rotor shaft with a single rigid disk with an unbalance mass. The rigid disk can be placed along position on the shaft. The rotor shaft is modelled and discretized into 11 Timoshenko beam elements with four degrees of freedom at each node as shown in Figure 5.1.

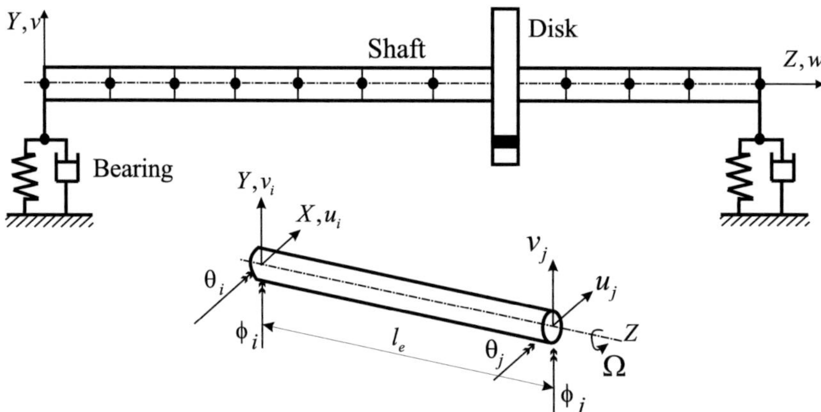

Figure 5.1: Rotor with a single rigid disk supported on both ends and coordinate system

The shaft element is modelled by 2-node beam elements, which have 8 degrees of freedom (2 translations and 2 rotations at each node), hence the dimension of the element matrix is 8 × 8.

- The displacements and rotations in horizontal and vertical direction are denoted by u, v, θ, ϕ, respectively. It is assumed that no displacements occur in the axial direction and u_i, v_i, θ_i, ϕ_i represent the displacement and rotation at node i, respectively.

- The shaft geometry is described by the length L and diameter d.

- The shaft material properties is defined by density ρ, modulus of elasticity E, shear modulus G and Poisson's ratio ν.

- The beam rotation about Z-axis with constant rotational speed Ω.

Model of the cracked shaft is shown in Figure 5.1. The nodal displacement of beam element is defined by

$$\mathbf{q} = \begin{bmatrix} u_i & \phi_i & v_i & \theta_i & u_j & \phi_j & v_j & \theta_j \end{bmatrix}^T \tag{5.1}$$

The shaft is represented as a beam with a circular cross-section and is characterized by kinetic energy and strain energy. The general formulation of the kinetic energy for the length $\mathrm{d}z$ of the element (Figure 5.2) [46], [75], [112], [27] and [39].

$$\mathrm{d}T = \left[\frac{1}{2}\rho A \left(\dot{u}^2 + \dot{v}^2 \right) + \frac{1}{2}\rho I_d \left(\dot{\theta}^2 + \dot{\phi}^2 \right) + \frac{1}{2}\rho I_p \Omega^2 + \rho I_p \Omega \dot{\phi} \theta \right] \mathrm{d}z \tag{5.2}$$

The first term of Eq.(5.2) is the expression for the kinetic energy of a beam in bending, the second term is the effect of rotational inertia and the third term is a constant due to constant rotational speed and has no influence on the equations. The last term represents the gyroscopic effect. I_p is the mass polar moment of inertia about the rotor axis and I_d is the diametral moment of inertia about any axis perpendicular to the rotor axis as shown in Figure 5.2.

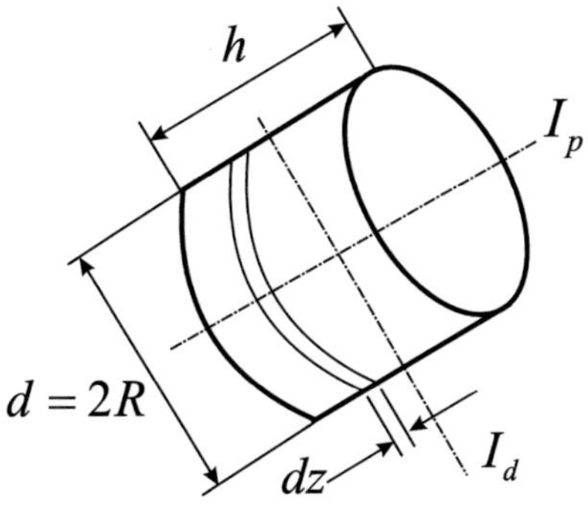

Figure 5.2: Geometry of cylindrical shaft element

$$I_p = \frac{1}{2}mR^2 \tag{5.3}$$

$$I_d = \frac{1}{12}m\left(3R^2 + h^2\right) \tag{5.4}$$

As the length $\mathrm{d}z$ ($h \cong 0$) is vanishingly small, it follows that

$$I = I_d = \frac{1}{2}I_p \tag{5.5}$$

The expression for the strain energy can be written as

$$\mathrm{d}U = \frac{1}{2}\left[EI\left\{\left(\frac{\mathrm{d}\theta}{\mathrm{d}z}\right)^2 + \left(\frac{\mathrm{d}\phi}{\mathrm{d}z}\right)^2\right\} + \frac{GA}{\kappa}\left\{\gamma_{xz}^2 + \gamma_{yz}^2\right\}\right]\mathrm{d}z \tag{5.6}$$

where E, G, A and κ are modulus of elasticity, shear modulus, shaft cross section area and shear coefficient, respectively. The shear deformations γ are related to the displacements and rotations through the following relations

$$\gamma_{xz} = -\phi - \frac{\mathrm{d}x}{\mathrm{d}z} \tag{5.7}$$

$$\gamma_{yz} = -\theta - \frac{\mathrm{d}y}{\mathrm{d}z} \tag{5.8}$$

The displacement x, y and rotations θ, ϕ of each node of the element can be written as functions of the displacements at the nodes, such that

$$\{q_x\} = \{u_1 \quad -\theta_1 \quad u_2 \quad -\theta_2\}^T \tag{5.9}$$

$$\{q_y\} = \{v_1 \quad -\phi_1 \quad v_2 \quad -\phi_2\}^T \tag{5.10}$$

By the expressions

$$x = [N_1]\{q_x\} \tag{5.11}$$

$$y = [N_1]\{q_y\} \tag{5.12}$$

$$\theta = -[N_2]\{q_y\} \tag{5.13}$$

$$\phi = [N_2]\{q_x\} \tag{5.14}$$

Here, a cubic shape function is proposed for y and a quadratic shape function is proposed for θ. The shape function coefficients are determined by requiring them to exactly satisfy both of the constraints. The resulting explicit form of two shape functions $[N_1]$ and $[N_2]$ are given in [1], [46]

$$[N_1] = \frac{1}{1+\vartheta} \begin{bmatrix} 1 + \vartheta\,(1-\varsigma) - 3\varsigma^2 + 2\varsigma^3 \\ l_e\left(\varsigma + \frac{1}{2}\vartheta\left(\varsigma - \varsigma^2\right) - 2\varsigma^2 + \varsigma^3\right) \\ 3\varsigma^2 - 2\varsigma^3 + \vartheta\varsigma \\ l_e\left(-\frac{1}{2}\vartheta\left(\varsigma - \varsigma^2\right) - \varsigma^2 + \varsigma^3\right) \end{bmatrix}^T \tag{5.15}$$

$$[N_2] = \frac{1}{(1+\vartheta)\,l_e} \begin{bmatrix} 6\left(-\varsigma + \varsigma^2\right) \\ l_e\left(1 - 4\varsigma + 3\varsigma^2 + \vartheta\,(1-\varsigma)\right) \\ -6\left(-\varsigma + \varsigma^2\right) \\ l_e\left(-2\varsigma + 3\varsigma^2 + \vartheta\varsigma\right) \end{bmatrix}^T \tag{5.16}$$

where l_e is the length of element and ς the nondimensional z coordinate defined as

$$\varsigma = \frac{z}{l_e} \tag{5.17}$$

ϑ is the ratio of the beam bending stiffness to the shear stiffness given by

$$\vartheta = \frac{12}{l_e^2}\left(\frac{EI}{\kappa GA}\right) = \frac{24}{l_e^2}\left(\frac{I}{\kappa A}\right)(1+\nu) \tag{5.18}$$

where ν is Poisson's ratio.

The integration of Eqs.(5.2) and (5.6) yields directly the kinetic and strain energy of the element, respectively

$$\begin{aligned}
T = {} & \frac{1}{2}\rho A I \left[\int_0^1 \{\dot{q}_x\}^T [N_1]^T [N_1] \{\dot{q}_x\}\,d\varsigma + \int_0^1 \{\dot{q}_y\}^T [N_1]^T [N_1] \{\dot{q}_y\}\,d\varsigma\right] + \\
& \frac{1}{2}\rho I l_e \left[\int_0^1 \{\dot{q}_x\} [N_2]^T [N_1] \{\dot{q}_x\}\,d\varsigma + \int_0^1 \{\dot{q}_y\}^T [N_2]^T [N_2] \{\dot{q}_y\}\,d\varsigma\right] + \\
& \rho I \Omega^2 l_e - 2\rho I \Omega l_e \int_0^1 \{\dot{q}_y\}^T [N_2]^T [N_2] \{\dot{q}_x\}\,d\varsigma
\end{aligned} \tag{5.19}$$

$$\begin{aligned}
U = {} & \frac{EI}{2l_e}\left[\int_0^1 \{q_x\}^T \frac{d}{d\varsigma}[N_2]^T \frac{d}{d\varsigma}[N_2]\{q_x\}\,d\varsigma + \int_0^1 \{q_y\}^T \frac{d}{d\varsigma}[N_2]^T \frac{d}{d\varsigma}[N_2]\{q_y\}\,d\varsigma\right] + \\
& \frac{EI}{2l_e}\left[\frac{12}{\vartheta}\int_0^1 \{q_x\}^T [N_3]^T [N_3]\{q_x\}\,d\varsigma + \int_0^1 \{q_y\}^T [N_3]^T [N_3]\{q_y\}\,d\varsigma\right]
\end{aligned} \tag{5.20}$$

where the shape function $[N_3]$ is defined as

$$[N_3] = [N_2] - \frac{1}{l_e}\frac{d}{d\varsigma}[N_1] \tag{5.21}$$

The equation of motion of the i-th element can be obtained from Lagrange equation

$$\frac{d}{dt}\left(\frac{\partial\,(T-U)}{\partial \dot{q}_i}\right) - \frac{\partial}{\partial q_i}(T-U) = 0 \tag{5.22}$$

Introducing the complex coordinates

$$\{q\} = \{q_x\} + j\,\{q_y\} \tag{5.23}$$

and differentiating Eq.(5.19) and Eq.(5.20) using Lagrange equations yields

$$
\begin{aligned}
0 = {} & \rho A I \int_0^1 [N_1]^T [N_1] \{\ddot{q}\}\,\mathrm{d}\varsigma + \rho I l_e \int_0^1 [N_2]^T [N_2] \{\ddot{q}\}\,\mathrm{d}\varsigma - 2j\rho I l_e \Omega \int_0^1 [N_2]^T [N_2] \{\dot{q}\}\,\mathrm{d}\varsigma \\
& + \frac{EI}{l_e}\left[\int_0^1 \left(\frac{\mathrm{d}}{\mathrm{d}\varsigma}[N_2]^T \frac{\mathrm{d}}{\mathrm{d}\varsigma}[N_2] \right)\mathrm{d}\varsigma + \frac{12}{\vartheta} \int_0^1 \left([N_3]^T [N_3] \right)\mathrm{d}\varsigma \right]\{q\}
\end{aligned}
\tag{5.24}
$$

The equation of motion can be written in matrix form

$$\left([M_T] + [M_R] \right)\{\ddot{q}\} - j\Omega\,[G]\,\{\dot{q}\} + [K]\,\{q\} = 0 \tag{5.25}$$

The expressions of the consistent matrices appearing in Eq.(5.25) can be obtained by performing the integrals of Eq.(5.24), which yields

- Translational inertia matrix $\mathbf{m}_T$

$$
\mathbf{m}_T = [M_T] = \frac{\rho A l_e}{420\,(1+\vartheta)^2}
\begin{bmatrix}
m_1 & l_e m_2 & m_3 & -l_e m_4 \\
 & l_e^2 m_5 & l_e m_4 & -l_e^2 m_6 \\
 & & m_1 & -l_e m_2 \\
sym & & & l_e^2 m_5
\end{bmatrix}
\tag{5.26}
$$

$$m_1 = 156 + 294\vartheta + 140\vartheta^2 \tag{5.27}$$

$$m_2 = 22 + 38.5\vartheta + 17.5\vartheta^2 \tag{5.28}$$

$$m_3 = 54 + 126\vartheta + 70\vartheta^2 \tag{5.29}$$

$$m_4 = 13 + 31.5\vartheta + 17.5\vartheta^2 \tag{5.30}$$

$$m_5 = 4 + 7\vartheta + 3.5\vartheta^2 \tag{5.31}$$

$$m_6 = 3 + 7\vartheta + 3.5\vartheta^2 \tag{5.32}$$

- Rotational inertia matrix $\mathbf{m}_R$

$$
\mathbf{m}_R = [M_R] = \frac{\rho I}{30 l_e\,(1+\vartheta)^2}
\begin{bmatrix}
m_7 & l_e m_8 & -m_7 & l_e m_8 \\
 & l_e^2 m_9 & -l_e m_8 & -l_e^2 m_{10} \\
 & & m_7 & -l_e m_8 \\
sym & & & l_e^2 m_9
\end{bmatrix}
\tag{5.33}
$$

$$m_7 = 36 \tag{5.34}$$

$$m_8 = 3 - 15\vartheta \tag{5.35}$$

$$m_9 = 4 + 5\vartheta + 10\vartheta^2 \tag{5.36}$$

$$m_{10} = 1 + 5\vartheta - 5\vartheta^2 \tag{5.37}$$

- Gyroscopic matrix **g**

$$\mathbf{g} = [G] = \frac{\rho I_p}{30 l_e \left(1 + \vartheta\right)^2} \begin{bmatrix} m_7 & l_e m_8 & -m_7 & l_e m_8 \\ & l_e^2 m_9 & -l_e m_8 & -l_e^2 m_{10} \\ & & m_7 & -l_e m_8 \\ sym & & & l_e^2 m_9 \end{bmatrix} \tag{5.38}$$

Because $I = I_d = \frac{1}{2} I_p$ (Eq. (5.5)), then

$$\mathbf{g} = 2\mathbf{m}_R = [G] = \frac{\rho I}{15 l_e \left(1 + \vartheta\right)^2} \begin{bmatrix} m_7 & l_e m_8 & -m_7 & l_e m_8 \\ & l_e^2 m_9 & -l_e m_8 & -l_e^2 m_{10} \\ & & m_7 & -l_e m_8 \\ sym & & & l_e^2 m_9 \end{bmatrix} \tag{5.39}$$

- Stiffness matrix **k**

$$\mathbf{k} = [K] = \frac{EI}{l_e^3 \left(1 + \vartheta\right)} \begin{bmatrix} 12 & 6 l_e & -12 & 6 l_e \\ & \left(4 + \vartheta\right) l_e^2 & -6 l_e & \left(2 - \vartheta\right) l_e^2 \\ & & 12 & -6 l_e \\ sym & & & \left(4 + \vartheta\right) l_e^2 \end{bmatrix} \tag{5.40}$$

Damping is assumed to be proportional to the mass and stiffness matrix (Rayleigh damping) and expressed by a damping factors α_c and β_c

$$\mathbf{c} = \alpha_c \left(\mathbf{m}_T + \mathbf{m}_R\right) + \beta_c \mathbf{k} \tag{5.41}$$

α_c and β_c are identified from considering the damping ratio versus frequency.

Hence, the equation of motion of the complete uncracked shaft model can be written as

$$\left(\mathbf{m}_T + \mathbf{m}_R\right) \ddot{\mathbf{q}} + \left(\mathbf{c} + \Omega \mathbf{g}\right) \dot{\mathbf{q}} + \mathbf{k} \mathbf{q} = \mathbf{0} \tag{5.42}$$

5.1.2 Disk model

The rigid disk has four degrees of freedom and the generalized coordinates are two translations of the mass centre and two rotations of the plane of the disk. The governing equations for the disk are derived in a similar fashion as the shaft element, therefore its nodal displacement vector is $[u \ \phi \ v \ \theta]^T$. If the rotor has a rigid disk, the mass centre of each rigid disk is located at the node that is shared between two elements of the shaft as shown in Figure 5.3.

If the disk is placed at node i of mass m_d, thickness L_d, outer radius R_{out-d}, and inner radius R_{in-d} as shown in Figure 5.3, then the mass polar moments of inertia about the rotor axis is [3]

$$I_{p-disk} = \frac{1}{2} m_d \left(R_{out-d}^2 + R_{in-d}^2\right) \tag{5.43}$$

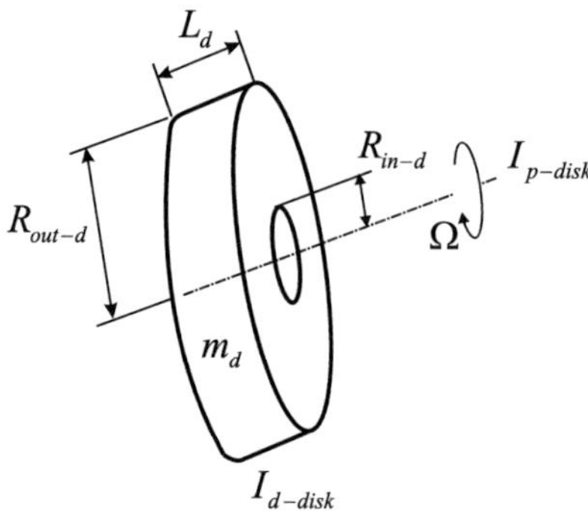

Figure 5.3: Geometry of rigid disk

and the diametral moment of inertia about any axis perpendicular to the rotor axis is given by

$$I_{d-disk} = \frac{1}{12}m_d \left(3R_{out-d}^2 + 3R_{in-d}^2 + L_d^2\right) \tag{5.44}$$

Hence, the equations of motion of the disk at node i are given by

$$\mathbf{M}_d\ddot{\mathbf{q}}_i + \mathbf{G}_d\dot{\mathbf{q}}_i = \mathbf{0} \tag{5.45}$$

$$\begin{bmatrix} m_d & 0 & 0 & 0 \\ 0 & I_{d-disk} & 0 & 0 \\ 0 & 0 & m_d & 0 \\ 0 & 0 & 0 & I_{d-disk} \end{bmatrix} \begin{Bmatrix} \ddot{u} \\ \ddot{\phi} \\ \ddot{v} \\ \ddot{\theta} \end{Bmatrix} + \Omega \begin{bmatrix} 0 & 0 & 0 & 0 \\ 0 & 0 & 0 & -I_{p-disk} \\ 0 & 0 & 0 & 0 \\ 0 & I_{p-disk} & 0 & 0 \end{bmatrix} \begin{Bmatrix} \dot{u} \\ \dot{\phi} \\ \dot{v} \\ \dot{\theta} \end{Bmatrix} = \begin{Bmatrix} 0 \\ 0 \\ 0 \\ 0 \end{Bmatrix} \tag{5.46}$$

where the first and second matrices are the mass matrix and the gyroscopic matrix of the disk, respectively and Ω is the shaft rotational speed. Since the disk is assumed to be rigid, the stiffness matrix of the disk vanishes.

The expression of the mass matrix which is the summation of the translational $\mathbf{m}_T$ and rotational mass matrices $\mathbf{m}_R$, the stiffness matrix $\mathbf{k}$, the external damping matrix which is assumed as proportional damping $\mathbf{c} = \alpha_c\mathbf{k}$ (the internal damping has been neglected) and the skew-symmetric gyroscopic matrix $\mathbf{g}$ is written in Eq.(5.39). The mass and gyroscopic matrices for the rigid disk are given by Eq.(5.45). Hence, the FE equation of motion of the shaft element including rigid disk is given by

$$\mathbf{M}\ddot{\mathbf{q}} + (\mathbf{C} + \Omega\mathbf{G})\dot{\mathbf{q}} + \mathbf{K}\mathbf{q} = \mathbf{F}^{ub} + \mathbf{F}^g \tag{5.47}$$

where $\mathbf{M}$ and $\mathbf{G}$ are the mass and gyroscopic matrices of the shaft and rigid disk. $\mathbf{C}$ and $\mathbf{K}$ are the external damping and stiffness matrices of the shaft. $\mathbf{F}^{ub}$ and $\mathbf{F}^g$ define the vector of unbalance force and gravity force for the degree of freedom $[u \ \phi \ v \ \theta]^T$ and is given by

$$\mathbf{F}^{ub} = \begin{bmatrix} m_{ub}\varepsilon\Omega^2 \cos\left(\Omega t + \delta\right) \\ 0 \\ m_{ub}\varepsilon\Omega^2 \sin\left(\Omega t + \delta\right) \\ 0 \end{bmatrix} \tag{5.48}$$

$$\mathbf{F}^{g} = \begin{bmatrix} 0 \\ 0 \\ -mg \\ 0 \end{bmatrix} \tag{5.49}$$

where m_{ub} and ε are mass unbalance and eccentricity, respectively. δ defines the initial angular position with respect to the Z-axis and m is the mass for each element of the shaft.

5.2 Model of the cracked shaft

5.2.1 Published models of cracked shaft

A FE model for the cracked rotor proposed by Sinou and Lees [136], [137], [135] is introduced. The numerical results of this model will be compared to the FE model of the cracked rotor using CZM. The transverse crack in the rotor system is modelled as a breathing crack that opens and closes in a synchronous manner as the shaft rotates. When the crack opens, there is a reduction in the cross sectional area moment of inertia of the element. The cracked element stiffness matrix $\mathbf{k}_{cr}$ based on Timoshenko beam theory is given by Sinou and Lees [136]. Note that ratio of crack depth to the shaft radius $\mu = a/R$ is used where a defines the crack depth and R the shaft radius. Furthermore, A_{ucr} defines the rest of uncracked area of the cross section and $\tilde{d}$ is the distance from the axis X to the centroid of the cross section (Figure 5.4) where $\tilde{I}_X$ and $\tilde{I}_Y$ are the area moments of inertia about the X and Y axes, respectively.

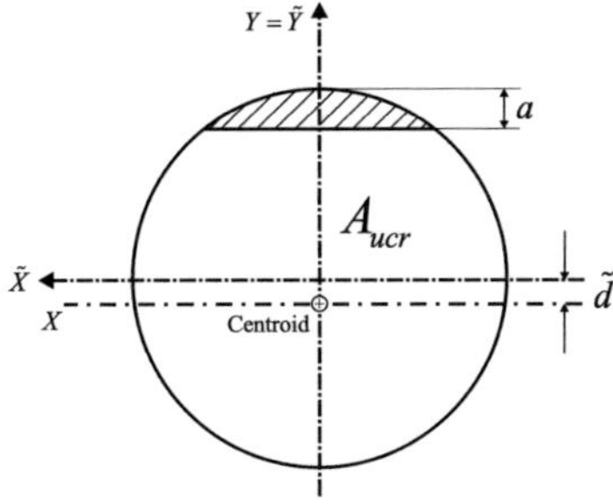

Figure 5.4: Cross section of the cracked shaft

$$\tilde{I}_Y = \frac{1}{4}R^4 \left[(1-\mu)(1-4\mu+2\mu^2)\sqrt{2\mu-\mu^2} + \arccos(1-\mu) \right] \tag{5.50}$$

$$\tilde{I}_X = \frac{\pi}{4}R^4 + R^4 \left[\frac{2}{3}(1-\mu)(2\mu-\mu^2)^{\frac{3}{2}} + \frac{1}{4}(1-\mu)(1-4\mu+2\mu^2)\sqrt{2\mu-\mu^2} \right]$$
$$- R^4 \left[\frac{1}{4}\arcsin\sqrt{2\mu-\mu^2} \right] \tag{5.51}$$

$$A_{ucr} = R^2 \left[(1-\mu)\sqrt{2\mu-\mu^2} + \arccos(1-\mu) \right] \tag{5.52}$$

$$\tilde{d} = \frac{2R^3}{3A_{ucr}}\left(2\mu-\mu^2\right)^{\frac{3}{2}} \tag{5.53}$$

The area moments of inertia of the cracked element cross-section about its centroidal axes are written as

$$I_X = \tilde{I}_X - A_{ucr}\tilde{d}^2 \tag{5.54}$$

$$I_Y = \tilde{I}_Y \tag{5.55}$$

Hence, the stiffness matrix due to the transverse crack $\mathbf{k}_{cr}$ can be obtained at the crack location

$$\mathbf{k}_{cr} = \frac{E}{L^3}\begin{bmatrix} 12I_x & 6LI_x & 0 & 0 & -12I_x & 6LI_x & 0 & 0 \\ 6LI_x & (4+\vartheta)L^2I_x & 0 & 0 & -6LI_x & (2-\vartheta)L^2I_x & 0 & 0 \\ 0 & 0 & 12I_y & -6LI_y & 0 & 0 & -12I_y & -6LI_y \\ 0 & 0 & -6LI_y & (4+\vartheta)L^2I_y & 0 & 0 & 6LI_y & (2-\vartheta)L^2I_y \\ -12I_x & -6LI_x & 0 & 0 & 12I_x & -6LI_x & 0 & 0 \\ 6LI_x & (2-\vartheta)L^2I_x & 0 & 0 & -6LI_x & (4+\vartheta)L^2I_x & 0 & 0 \\ 0 & 0 & -12I_y & 6LI_y & 0 & 0 & 12I_y & 6LI_y \\ 0 & 0 & -6LI_y & (2-\vartheta)L^2I_y & 0 & 0 & 6LI_y & (4+\vartheta)L^2I_y \end{bmatrix} \tag{5.56}$$

After assembling the different shaft elements, adding the element of the rigid disk and the bearing forces, the complete cracked rotor equation is given by

$$\mathbf{M\ddot{q}} + (\mathbf{C} + \Omega\mathbf{G})\,\dot{\mathbf{q}} + (\mathbf{K} - f(\Omega t)\mathbf{K}_{cr})\,\mathbf{q} = \mathbf{F^{ub}} + \mathbf{F^g} \tag{5.57}$$

The assembling schema of the complete cracked rotor system is displayed in Figure 5.5. The deflection of the slowly rotating shaft is softly and continuously changing with Ωt. In case of a small crack ($a/d \leq 0.20$), the crack opens and closes rather abruptly. The transition is less abrupt in case of deeper cracks. Mayes and Davis [90] suggested that the continuous breathing steering function $f(\Omega t)$ for deeper cracks, because it comes closer to reality. This usual model has been adopted by many researchers [3], [2] where the opening and closing of the crack is described by a cosine function by assuming that the gravity force is much greater than the unbalance force and given by

$$f(\Omega t) = \frac{1}{2}\left(1 \pm \cos(\Omega t)\right) \tag{5.58}$$

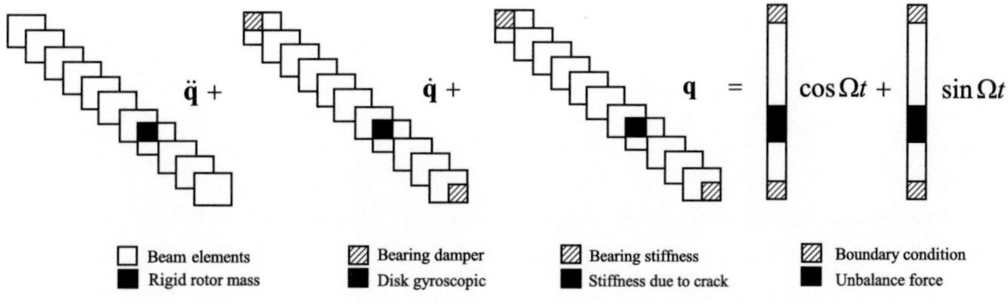

Figure 5.5: Assembling of the complete cracked rotor system

The plus sign of the cosine term in this function is used when the crack is fully open and symmetric with the negative Y-axis at $t{=}0$ while the negative sign is used when the crack is fully open and symmetric with the positive Y-axis at $t{=}0$. The sign change of the cosine term only rotates the whirl orbit by π radian without affecting its shape. The deflection of the slowly rotating shaft is softly and continuously changing with Ωt. The transition is less abrupt in case of deeper cracks.

5.2.2 Model based on asymmetric area moments of inertia

Here, the crack stiffness element is modelled using the CZM. Two FE one dimensional continuum rotor models using CZM are proposed:

1. Model based on asymmetric area moments of inertia due to the crack is modelled by the breathing steering function obtained in Chapter 3 instead of Mayes' model.

2. Model based on the direct TSL using one zero thickness element, which is placed between the continuum elements. This model is discussed in the next section.

If there is a crack in element-j, the new cracked element stiffness matrix $\mathbf{k}^{j}_{ce}$ can be expressed based on the stiffness matrix due to the transverse crack $\mathbf{k}_{cr}$ at the crack location (Eq.(5.56)) and is written

$$\mathbf{k}^{j}_{ce} = f_{coh}\left(\Omega t\right)\mathbf{k}^{j}_{cr} \tag{5.59}$$

where $\mathbf{k}^{j}$ is the 8×8 element stiffness matrix when the crack is fully closed (Eq.(5.40)) and $f_{coh}\left(\Omega t\right)$ expresses the breathing steering function modelled by CZM obtained by curve fitting (Eqs.(3.90-91)).

In case rotor without disk, the equation of motion yields

$$\mathbf{M\ddot{q}} + \mathbf{C\dot{q}} + \mathbf{K}_{c}\mathbf{q} = \mathbf{0} \tag{5.60}$$

where the global stiffness matrix is

$$\mathbf{K}_c = \begin{bmatrix} \cdots & \cdots & \cdots \\ \cdots & \mathbf{k}_{ce}^{j} & \cdots \\ \cdots & & \cdots \end{bmatrix}$$

(5.61)

The breathing crack is included in the model, i.e. when the equivalent cracked beam having a reduced cross section and a suitable length is inserted in the FE model of the rotor, the stiffness matrix will have variable values in the 8×8 elements that correspond to the cracked beam element, which may vary between a maximum stiffness (corresponding to closed crack) and a minimum stiffness (corresponding to open crack), instead of constant values. After assembling the different shaft elements, the first proposed FE model of the cracked rotor using CZM is shown in Figure 5.6.

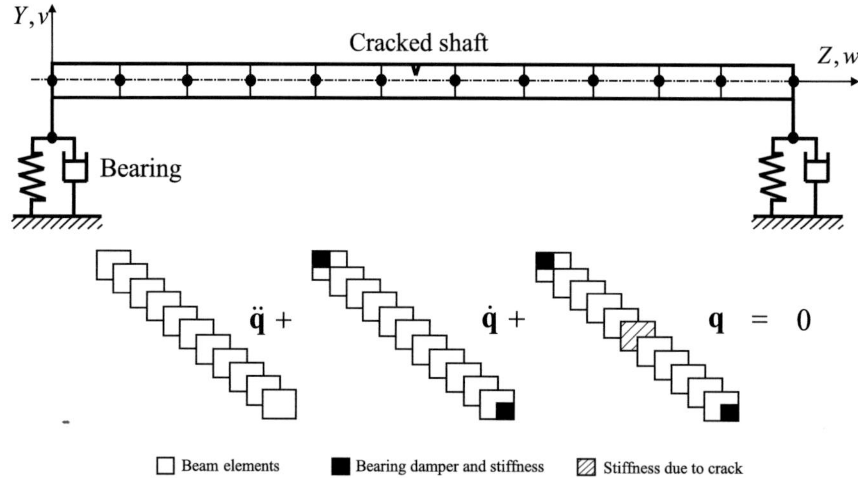

Figure 5.6: Assembling of the cracked rotor system: first proposed FE model

Eq.(5.60) can be rewritten in the form of a state equation

$$\mathbf{p} = \dot{\mathbf{q}}$$

(5.62)

$$\dot{\mathbf{p}} = \ddot{\mathbf{q}}$$

(5.63)

Substitution Eqs.(5.62-63) into Eq.(5.60) yields

$$\dot{\mathbf{p}} = -\mathbf{M}^{-1}\mathbf{C}\mathbf{p} - \mathbf{M}^{-1}\mathbf{K}\mathbf{q}$$

(5.64)

Introducing

$$\mathbf{r} = \left\{ \begin{array}{c} \mathbf{q} \\ \mathbf{p} \end{array} \right\}$$

(5.65)

Thus, the state equation of the cracked rotor is

$$\dot{\mathbf{r}} = \mathbf{A}\mathbf{r} \tag{5.66}$$

where $\mathbf{A}$ is the system matrix and is defined by

$$\mathbf{A} = \begin{bmatrix} \mathbf{0} & \mathbf{I} \\ \mathbf{M}^{-1}\mathbf{K} & -\mathbf{M}^{-1}\mathbf{C} \end{bmatrix} \tag{5.67}$$

Eq.(5.66) can be solved by any numerical procedures to obtain the response of the system [109], [152].

5.2.3 Finite element results

Only the static bending moment is considered, because the bending moment due to the inertia force distribution associated to the rotor vibration usually gives a small contribution with respect to the static bending moment (often only near the rotor critical speeds it can become significant). This assumption is generally acceptable for heavy horizontal rotating machines. Hence, the linear approach for this case is suitably accurate.

5.2.3.1 Natural frequencies

Based on the assembling of the complete cracked rotor system as shown in Figure 5.5 and Figure 5.6, the lower eigenfrequencies of the breathing cracked shaft can be determined. In Table 5.1, the physical and geometrical parameters of the cracked shaft is listed.

Table 5.1: Shaft parameters

Symbol	Parameter	Value
d	Diameter of the rotor shaft	$0.08\,\mathrm{m}$
L	Length of the rotor shaft	$1.0\,\mathrm{m}$
a	Crack depth	$0.008\,\mathrm{m}$
E	Modulus of elasticity	$210\,\mathrm{GPa}$
ν	Poisson ratio	0.3
ρ	Density	$7\,850\,\mathrm{kg/m^3}$
A	Cross section area	$0.005\,03\,\mathrm{m^2}$
I	Area moment of inertia	$2.01 \times 10^{-6}\,\mathrm{m^4}$
r_0	Radius of gyration	$0.020\,\mathrm{m}$
m_s	Mass of the shaft	$39.458\,4\,\mathrm{kg}$

Figure 5.7 shows the first three natural frequencies of the uncracked and cracked rotor without disk. Shifts are observed for the third natural frequency. In particular, the third natural frequency of the cracked rotor is significantly lower than those corresponding to the uncracked rotor. The absolute percentage differences for three frequencies are 1.62%, 0.38% and 3.74%, respectively.

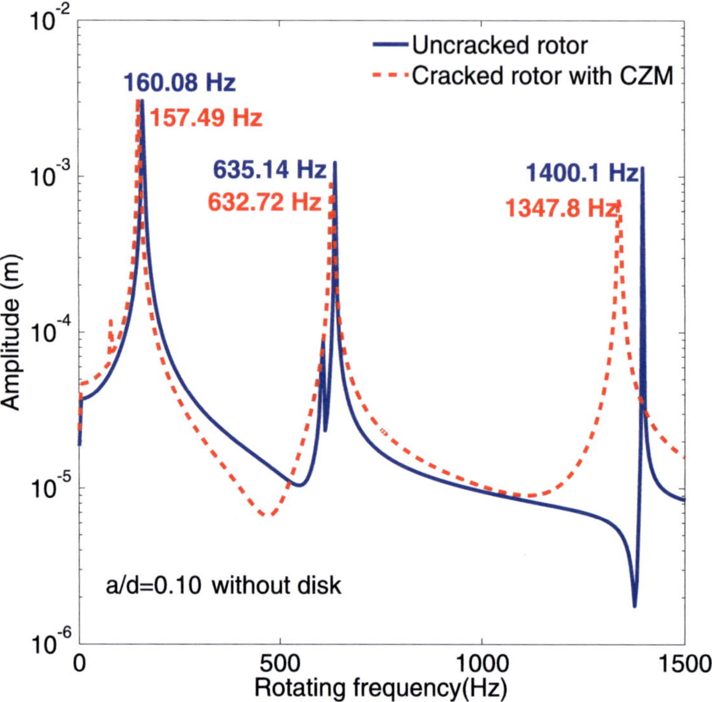

Figure 5.7: Natural frequencies for the uncracked and cracked rotor without disk

5.2.3.2 Mode shapes

Figure 5.8 illustrates that mode shapes for the first three natural frequencies of the cracked and uncracked rotor without disk are rather similar. Shifts are observed for the third natural frequency. In particular the crack depth does not affect significantly the first and second natural frequencies of the cracked rotor. However, the crack depth affects slightly the third natural frequency.

5.2.3.3 Effect of crack depth

Figure 5.9 displays the ratio of the first natural frequency of the cracked rotor to the first natural frequency of the corresponding uncracked rotor as a function of the relative crack depth a/d. The natural frequencies of the cracked rotor are lower than the natural frequencies of the corresponding uncracked rotor, as expected. The reduction in natural frequency is due to the local flexibility introduced by the crack and as a result that the stiffness is decreased. These differences increase with the depth of the crack which are obtained from Figure 5.10.

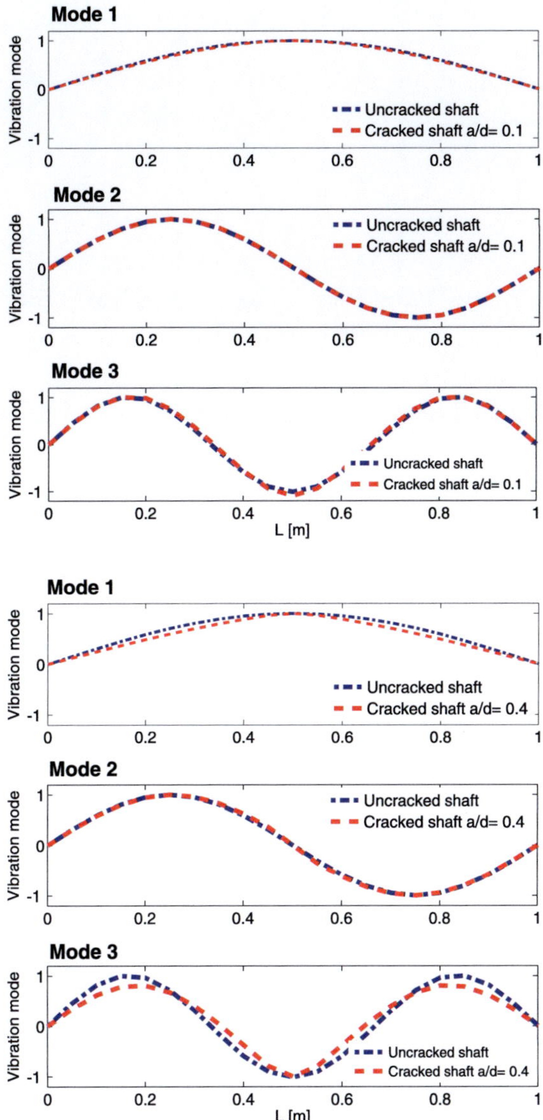

Figure 5.8: Variation of mode shapes of the uncracked and cracked rotor for the first three natural frequencies for relative crack depth a/d=0.1 and 0.4

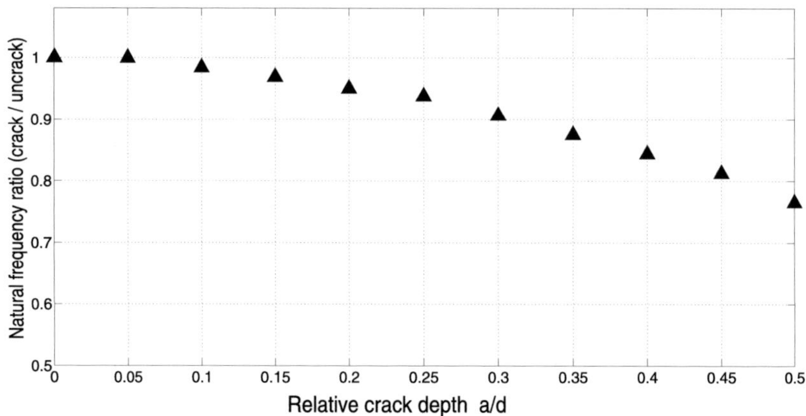

Figure 5.9: First natural frequency ratio for different crack depths

5.2.3.4 Comparison of results

The FE results in the previous section are compared with the analytical results based on the Timoshenko beam theory given in Appendix A.1. The Timoshenko beam theory can be used to derive a mathematical model for the open crack only and the results can be used as a rough estimation to validate the FE results. Using the eigenfrequencies of the undamaged shaft Eq.(A.29)

$$
\omega_i^2 = \frac{\frac{1}{EI}\left[\left(\frac{n\pi}{L}\right)^2\left(\frac{EI\rho}{\kappa G} + \rho A r_0^2\right) + \rho A\right]}{2\frac{\rho^2 A r_0^2}{EI\kappa G}}
$$
$$
\pm \frac{\sqrt{\frac{1}{(EI)^2}\left[\left(\frac{n\pi}{L}\right)^2\left(\frac{EI\rho}{\kappa G} + \rho A r_0^2\right) + \rho A\right]^2 - 4\frac{\rho^2 A r_0^2}{EI\kappa G}\left(\frac{n\pi}{L}\right)^4}}{2\frac{\rho^2 A r_0^2}{EI\kappa G}}
\tag{5.68}
$$

Substituting the shaft parameters listed in Table 5.1, the first three natural frequencies are

- $\omega_1 = 1\,015{,}9$ rad/s $= 161.7$ Hz

- $\omega_2 = 4\,005{,}8$ rad/s $= 637.5$ Hz

- $\omega_3 = 8\,810{,}9$ rad/s $= 1\,402.3$ Hz

In Table 5.2, a comparison between the results obtained from FE and Timoshenko beam theory is presented for the uncracked shaft.

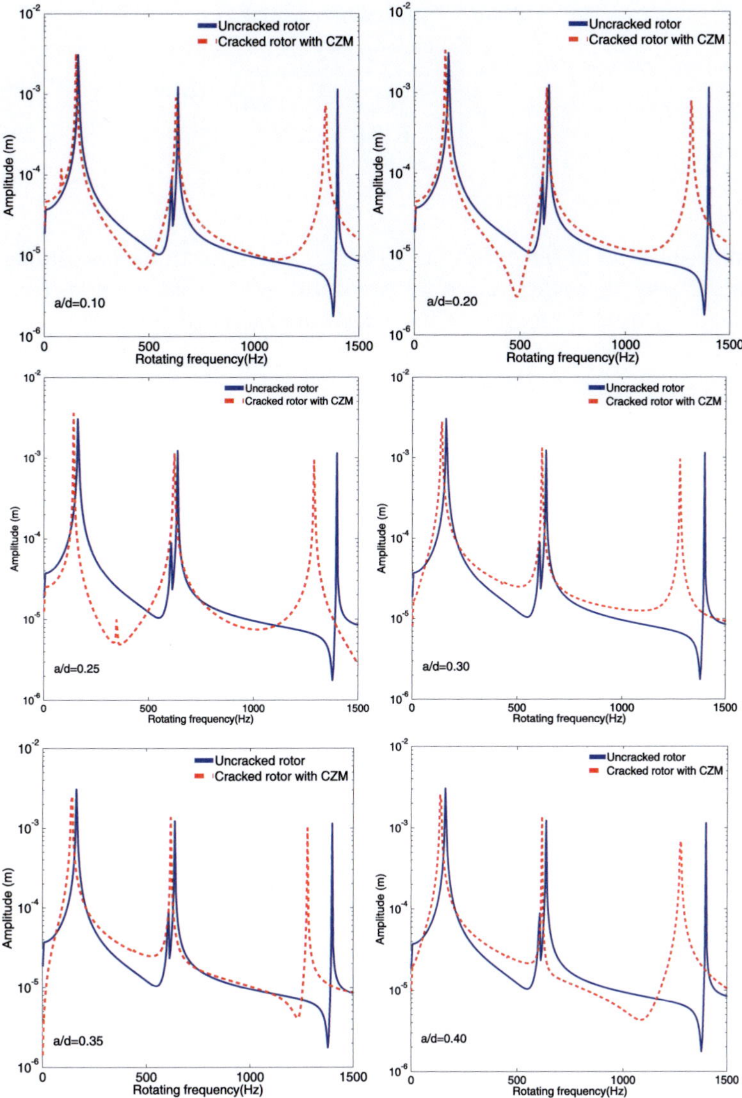

Figure 5.10: First of three natural frequencies for different crack depths of shaft without disk for relative crack depth a/d from 0.1 to 0.4

Table 5.2: Natural frequencies of uncracked shaft without disk at rest in [Hz]

Mode	Timoshenko beam-theory	FE
1	161.7	160.1
2	637.5	635.1
3	1 402.3	1 400.1

The natural frequencies of the open cracked shaft can be determined according to Mayes' model which relates the change in second moment of area to the relative crack depth a/d (Eq.(2.50) and Figure 2.32) for $a/d = 0.1$, substitute I_{cr} into I, we have

- $\omega_{cr_1} = 1\,000{,}6$ rad/s $= 159.3$ Hz

- $\omega_{cr_2} = 3\,806{,}1$ rad/s $= 605.8$ Hz

- $\omega_{cr_3} = 7\,967{,}0$ rad/s $=1\,268.0$ Hz

Table 5.3 summarizes the results for the natural frequencies of the cracked shaft using Timoshenko beam theory and FE.

Table 5.3: Natural frequencies of cracked shaft without disk at rest in [Hz]

Mode	Timoshenko beam-theory	FE
1	159.3	157.5
2	605.8	632.7
3	1 268.0	1347.8

5.2.4 Model based on zero-thickness element

In the second model, one element having zero-thickness is placed between the continuum elements. The stiffness of zero-thickness element is defined by the stiffness of cohesive element which is defined by TSL law according to the rest of uncracked area [84].

$$k_{coh}^{j} = K_p \left[A - (A - A_{ucr}) f(\Omega t) \right] \tag{5.69}$$

where $A = \frac{\pi}{4} d^2$ and A_{ucr} are area of cross section without crack and the remains uncracked area of the cross section (Eq.(5.52)), respectively. $f(\Omega t)$ expresses the continuous breathing steering function obtained from curve fitting and K_p the penalty stiffness which either can be obtained directly from the TSL

$$K_p = \frac{\sigma_n}{\delta_1} \tag{5.70}$$

or can be estimated from element stiffness. Here, $K_p = (10^{12} \div 5 \cdot 10^{13})$ N/m^3 is used that ensures a stiff connection between the surfaces of the material discontinuity. The penalty stiffness should be large enough to provide connection between the two elements but small enough to avoid numerical problem in a FE analysis. A first value of the penalty stiffness is obtained from estimation of the element stiffness as follows

$$K_p = \frac{48EI/l_{el}^3}{A_{el}} = \frac{48 \cdot 210 \cdot 10^9 \frac{\pi}{64} 0.08^4 / 0.1^3}{\frac{\pi}{4} 0.08^2} = 4.032 \cdot 10^{12} \text{ N/m}^3$$

where shaft is discretized into 10 elements, so that $l_{el} = 0.1L$.

The other estimate of the penalty stiffness is based on the TSL. The applied cohesive law has mechanical properties as discussed in Chapter 2 for steel with maximum traction $\sigma_n = 250$ MPa which is assumed to be the same as the yield strength of material. The maximum separation at the end of the elastic zone is assumed to be $\delta_1 \cong 10\mu m$. The TSL is shown in Figure 5.11.

From Eq.(5.69), it is clear that if the crack is fully closed said $f(\Omega t) = 0$ and area of the cross section is A, while if the crack is fully open $f(\Omega t) = 1$ and area of the cross section is the remaining of uncracked area of the cross section A_{ucr} (Figure 5.4). The material properties of the cohesive elements are assumed to be isotropic, i.e. the direct stiffness of cohesive element in x and y direction are the same. Further, the coupling stiffnesses are also assumed to be constant and same as the direct stiffnesses.

$$F_{coh}^{u} = k_{coh}^{u} u + k_{coh}^{uv} v \tag{5.71}$$

$$F_{coh}^{v} = k_{coh}^{v} u + k_{coh}^{uv} u \tag{5.72}$$

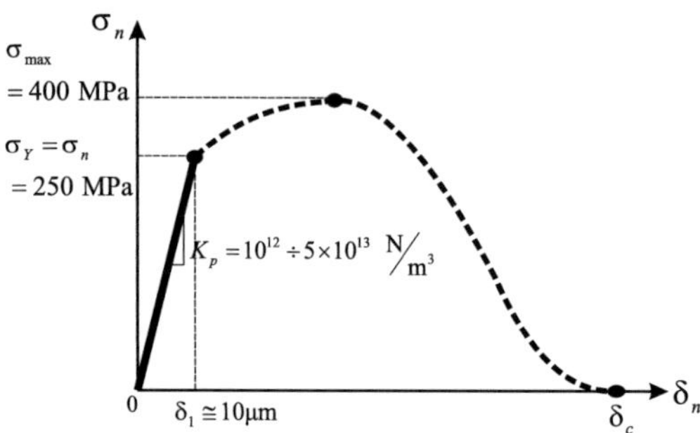

Figure 5.11: Traction-separation law in FE computation

Since $k_{coh}^u = k_{coh}^v = k_{coh}^{uv} = k_{coh}$, thus, the local element stiffness matrix at the cohesive crack is

$$\mathbf{k}_{coh}^j = \begin{bmatrix} k_{coh}^j & 0 & k_{coh}^j & 0 \\ 0 & 0 & 0 & 0 \\ k_{coh}^j & 0 & k_{coh}^j & 0 \\ 0 & 0 & 0 & 0 \end{bmatrix} \tag{5.73}$$

In case of a rotor without disk, the equation of motion yields

$$\mathbf{M}\ddot{\mathbf{q}} + \mathbf{C}\dot{\mathbf{q}} + \mathbf{K}_{coh}\mathbf{q} = \mathbf{0} \tag{5.74}$$

where $\mathbf{K}_{coh}$ is the global element stiffness matrix,

$$\mathbf{K}_{coh} = \begin{bmatrix} \cdots & \cdots & \cdots \\ \cdots & \mathbf{k}_{coh}^j & \cdots \\ \cdots & & \cdots \end{bmatrix} \tag{5.75}$$

The assembling schema for the FE cracked rotor model is shown in Figure 5.12.

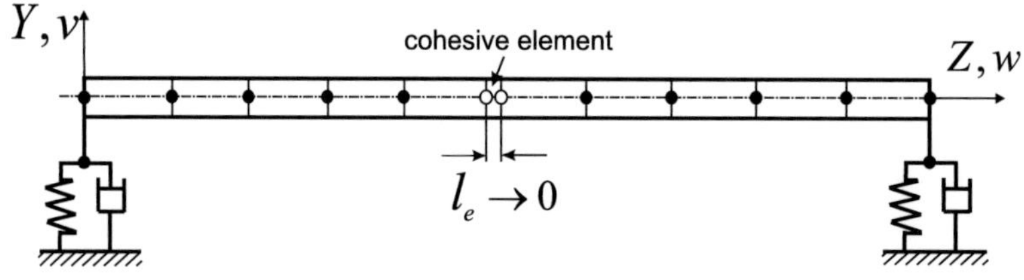

Figure 5.12: Assembling of the cracked rotor system: second proposed FE model

Table 5.4: First natural frequencies of cracked shaft without disk in [Hz]

	a/d=0.10	a/d=0.20	a/d=0.25	a/d=0.30	a/d=0.35	a/d=0.40
Mayes'model	159.1	151.3	147.7	145.1	140.1	137.5
1st FE Model of CZM	157.5	150.1	147.1	144.7	140.1	136.8
2nd FE Model of CZM	159.2	152.4	149.2	147.2	142.5	140.3

Table 5.5: Second natural frequencies of cracked shaft without disk in [Hz]

	a/d=0.10	a/d=0.20	a/d=0.25	a/d=0.30	a/d=0.35	a/d=0.40
Mayes'model	633.9	625.3	620.3	616.0	607.5	602.1
1st FE Model of CZM	632.7	625.1	620.8	615.9	606.2	602.1
2nd FE Model of CZM	630.2	624.8	620.2	615.3	606.1	602.3

Figure 5.13 and Tables 5.4-5.5 show the first two natural frequencies of the cracked rotor without disk based on Mayes' model and two proposed FE models using CZM. Shifts are observed for the second FE model, for which the first natural frequencies are always smaller than Mayes' model. This can be caused by the partial opening of the cohesive elements around the fracture process zone near the crack tip. In particular, the second natural frequency of two FEs of CZM are obviously in good agreement, although the amplitude of the second FE model is significantly lower than Mayes' model.

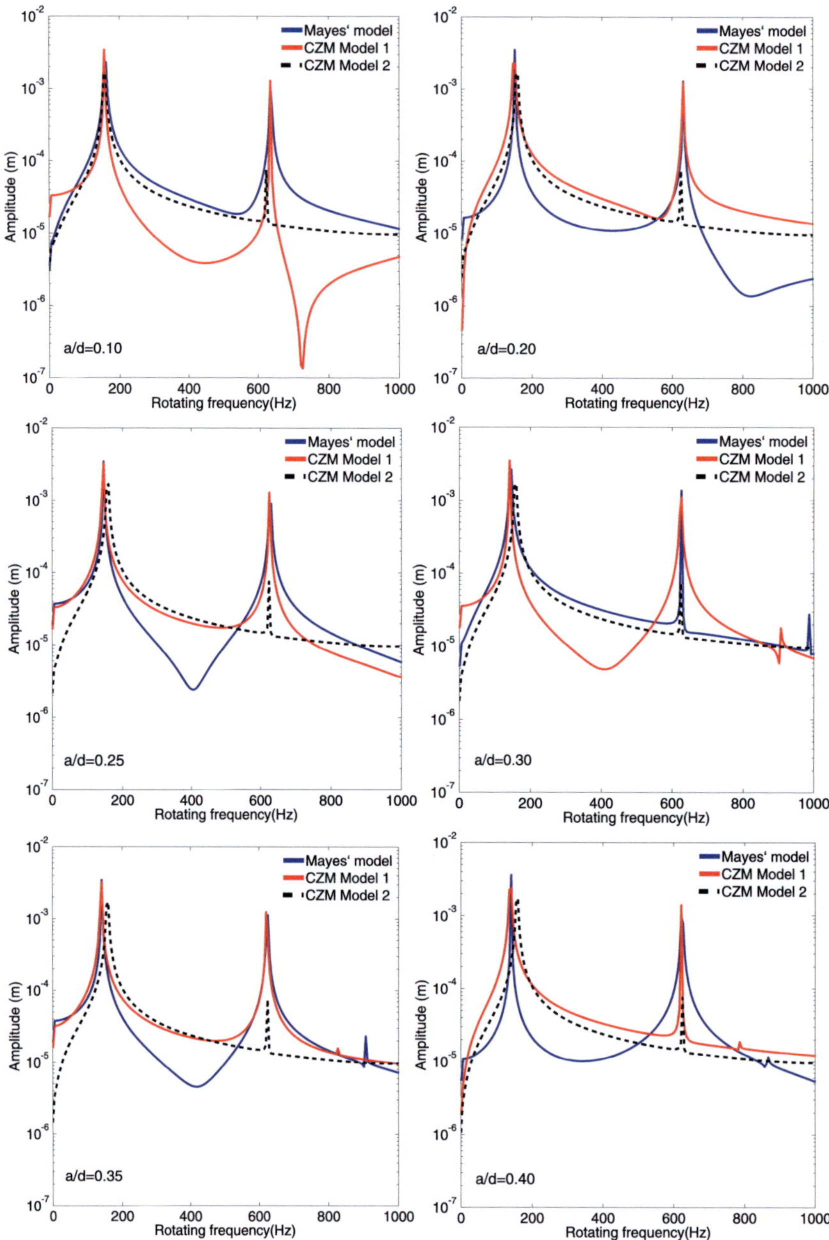

Figure 5.13: Natural frequencies for the cracked rotor without disk based on Mayes' model, and two proposed FE models

5.3 Cracked rotor supported by rigid bearings with disk

5.3.1 Finite element results

Based on the assembling of the complete cracked rotor system as shown in Figure 5.5, the first three natural frequencies of the uncracked and cracked rotor with disk are estimated. Here, the first proposed FE is used. The shaft and disk parameters are shown in Table 5.1 and Table 5.6, respectively.

Table 5.6: Disk parameters

Symbol	Parameter	Value
D_{out-d}	Outer diameter of the disk	0.15 m
D_{in-d}	Inner diameter of the disk	0.08 m
L_d	Length of the disk	0.10 m
E	Modulus of elasticity	210 GPa
ρ	Density	$7\,850$ kg/m^3
I_p	Mass polar moment of inertia	$0.035\,69$ kg.m^2
I_d	Diametric moment of inertia	$0.026\,20$ kg.m^2
m_d	Mass of disk	$9.926\,3$ kg
ε	Eccentricity of the unbalance	0.070 m
m_{ub}	Mass unbalance	0.05 kg

Figure 5.14 represents the vertical and horizontal amplitude as a function of rotating frequency for the cohesive cracked and the uncracked rotor with disk. The first three frequencies of the uncracked and cracked rotor with disk are rather close. The absolute percentage differences for three frequencies are 0.40%, 3.68% and 1.69%, respectively. In comparison to the frequencies of the uncracked and cracked rotor, the values of natural frequencies for rotor with disk are generally lower.

5.3.2 Comparison of results

Two approximations have been used to estimate the critical speed of rotor. Rayleigh's method and Dunkerley's equation are suitable for estimating the fundamental frequency by hand calculation and are presented in Appendix A.2 and A.3. In general, Rayleigh's method overestimates and Dunkerley's equation underestimates the natural frequency. Dunkerley advanced a method to approximate the fundamental frequency of multirotor systems. It gives good results if damping is negligible. In the following, the natural frequencies of the uncracked shaft obtained from the FE model are compared with the fundamental natural frequencies calculated by Dunkerley's equation and Rayleigh's method. Table 5.7 shows the result of the natural frequencies for a cracked rotor without disk. It can be seen that the FE result is between the lower natural frequency estimation and the higher one. Further, it is accurate to use 11 elements in order to estimate the fundamental of natural frequency.

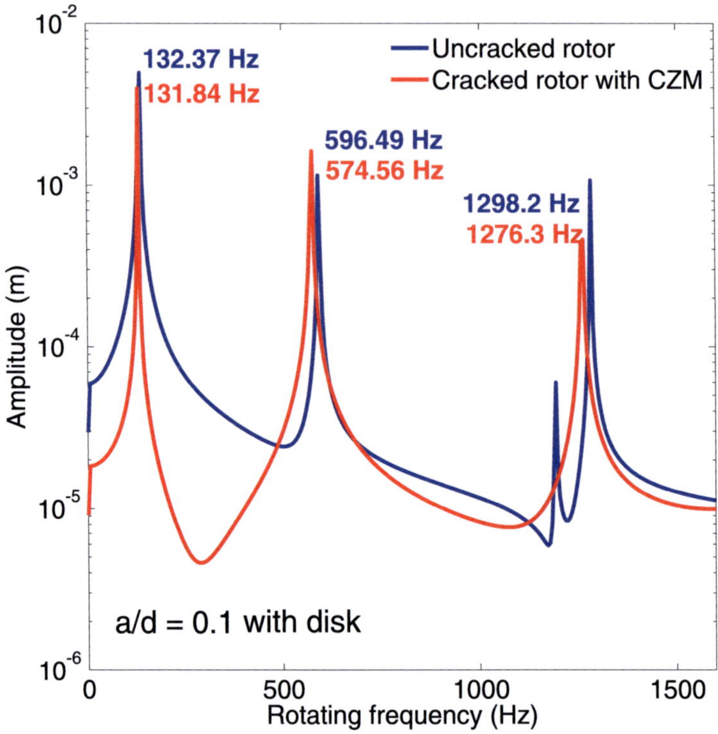

Figure 5.14: Natural frequencies for the uncracked and cracked rotor with disk

By using Dunkerley's equation Eq.(A.48), the lower bound of the first natural frequency of the undamaged shaft is

$$\omega_1^2 \geq \frac{EI}{L^3 \left(\frac{\rho AL}{97.417} + \frac{m_d}{48} \right)} \tag{5.76}$$

The natural frequency of uncracked shaft can be approximated and using Figure 2.32, for relative crack depth $a/d = 0.1$, $\Delta I/I = 0.003\,75$ or $I_{cr} = 2.003\,1 \times 10^{-6}$ m^4. Thus, the natural frequency of cracked shaft which is modelled by open cracked model are

$$\omega_{ucr} = \sqrt{\frac{1}{\left(\frac{7850 \cdot \frac{\pi}{4} 0.08^2 \cdot 1}{97.417} + \frac{9.926\,3}{48} \right)} \frac{210 \times 10^9 \cdot \frac{\pi}{64} 0.08^4}{1^3}} = 830.72 \text{ rad/s} \ (= 132.21 \text{ Hz})$$

$$\omega_{cr} = \sqrt{\frac{1}{\left(\frac{7850 \cdot \frac{\pi}{4} 0.08^2 \cdot 1}{97.417} + \frac{9.926\,3}{48} \right)} \frac{210 \times 10^9 \cdot 2.0031 \times 10^{-6}}{1^3}} = 829.16 \text{ rad/s} \ (= 131.96 \text{ Hz})$$

Table 5.7: Natural frequencies of uncracked shaft with disk in [Hz]

Mode	Dunkerley's equation	Rayleigh's method	FE
1	132.21	132.53	132.37

By using the Rayleigh's method Eq.(A.59), the upper bound of the first natural frequency of the undamaged shaft is

$$\omega_1^2 \leq \frac{\pi^4}{m_s + 2m_d} \frac{EI}{L^3} \tag{5.77}$$

The natural frequency of uncracked shaft can be approximated and using Figure 2.32, for relative crack depth $a/d = 0.1$, $\Delta I/I{=}0.003\,75$ or $I_{cr}{=}2.003\,1 \times 10^{-6}$ m^4. Thus, the natural frequency of cracked shaft which is modelled by open cracked model are

$$\omega_{ucr} = \sqrt{\frac{\frac{\pi}{4}}{39.4584 + 9.9263} \frac{210 \times 10^9 \cdot \frac{\pi}{64} 0.08^4}{1^3}} = 832.73 \text{ rad/s } (= 132.53 \text{ Hz})$$

$$\omega_{cr} = \sqrt{\frac{\frac{\pi}{4}}{39.4584 + 9.9263} \frac{210 \times 10^9 \cdot 2.0031 \times 10^{-6}}{1^3}} = 831.18 \text{ rad/s } (= 132.29 \text{ Hz})$$

Table 5.8 gives approximations for the natural frequency of the cracked rotor with disk. For the calculation of the fundamental natural frequency using Dunkerley's equation and Rayleigh's method, it is assumed that the crack is always open which leads to a reduction in the cross sectional area moment of inertia as described by Eq.(2.50). In comparison with the first natural frequency of the open crack model from both methods, the natural frequency for the breathing crack is slightly higher than the natural frequency estimated by Dunkerley's equation.

Table 5.8: Natural frequencies of cracked shaft with disk in [Hz]

Mode	Dunkerley's equation	Rayleigh's method	FE
1	131.96	132.29	131.84

The main results of the FE numerical investigation in this chapter are the following:

1. Two FE model approaches based on equivalent beam using CZM have been proposed. Comparison with the results in literature show good agreement, as long as a single crack with regular shape, i.e. rectilinear shape is considered.

2. Stiffness variation is defined by the function of the TSL corresponding to the stress acting in the crack. In the first FE model, breathing crack is modelled by a function of the angular position that is called the cohesive breathing steering function, obtained by curve fitting as discussed in Chapter 3. The second FE model implemented one zero thickness cohesive element which is placed between continuum elements. The continuous straight line for shallow crack and parabolic line in case of deep crack are used.

3. The second FE model is more realistic since the deflection line of a shaft with a crack in tension zone is given by the superposition of two parts: the deflection line of the uncracked shaft and the additional deflection caused by the local compliance of the crack. This additional part cannot be found from the elastic beam theory, because for the beam theory, a crack is the weakening of the bending stiffness on a zero length [42]. Therefore, using zero thickness cohesive element in FE is reasonable.

4. Disadvantage is the cost in terms of computation time, i.e. for FE model without CZM the computing time is about four times lower than FE model with CZM.

5. With realistic damping values the linear approach is suitably accurate and effect of unbalance is correctly predicted, even if the breathing mechanism is governed by the vibration.

6 Breathing crack simulation

The presence of a crack in a rotor reduces the stiffness of the system and the variable part of the rotor stiffness varies between a minimum (for open cracks) and a maximum (for closed cracks), depends on the so-called breathing mechanism. The breathing mechanism is known when the open and closed parts of the cracked area are known at all angular positions of the rotor. Non-linear behaviour of cracked rotor occurs when the breathing crack is not anymore determined by the static forces, but by the dynamic forces associated to the vibration or response of the system.

Generally the vibration response of the cracked rotor is small, so that the bending moment due to the dynamic forces (external forces and inertia forces) is smaller than the static bending moments due to the external forces (such as the weight and any other stationary force, in horizontal heavy rotors of industrial plants [11]). Therefore, the breathing mechanism is dominated by the static bending moment and the dependence of the stiffness variation on the vibration response can be neglected. Breathing mechanism with weight dominance during rotation is shown in Figure 6.1.

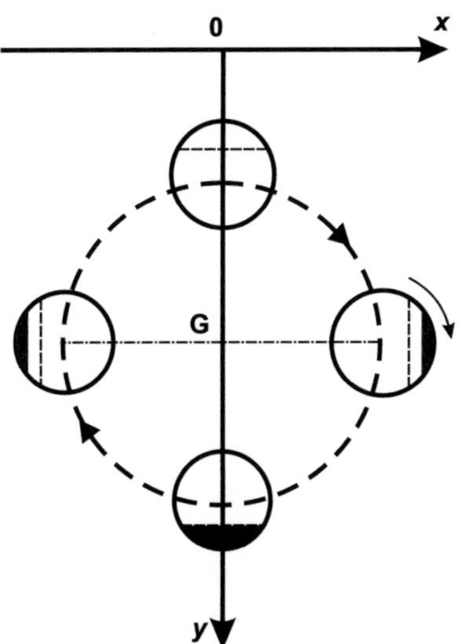

Figure 6.1: Breathing crack with weight dominance [21]

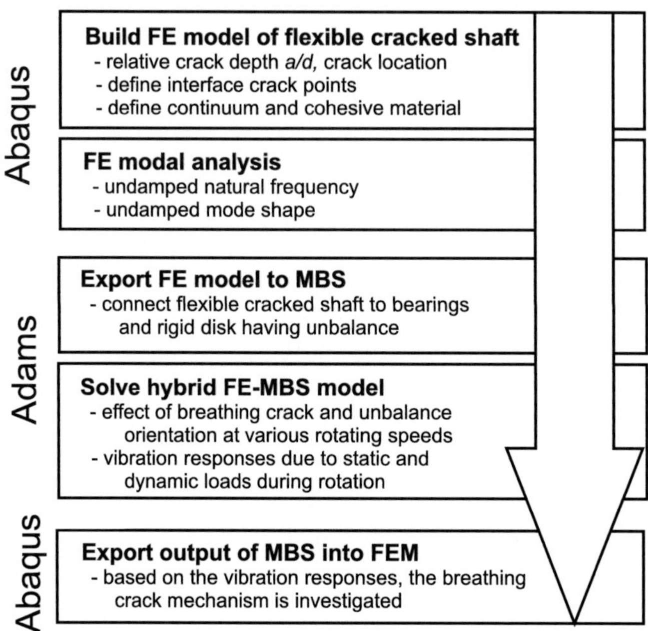

Figure 6.2: Principle of the simulation using an integration of FE and MBS [81]

In this chapter, the breathing mechanism of a cracked shaft de Laval rotor on rigid supports has been investigated. An integrated simulation process of finite element (FE) and multi-body simulation (MBS) is used. First, an elastic cracked shaft with various relative crack depths is modelled by FE software. The analysis deals with the natural frequencies and mode shapes of cracked and uncracked rotor. Furthermore, breathing crack under rotating load (non-rotating shaft) is also investigated. A bending load is applied and repeated for all different angular positions of the cracked shaft specimen. At the second step, the FE model of elastic cracked shafts is exported into MBS software in order to analyze the dynamic loads, due to the crack, unbalance and inertia force acting during rotation at different rotating speeds. The analyses consist of cracked shaft loaded by weight only and by weight and unbalance. The effect of orientation angle of the unbalance mass on the breathing crack behaviour has also been investigated. The case where an unbalance is located on the crack side, and the other case where the unbalance is located on the opposite side of the crack have been studied. In this work, three combinations of unbalance mass and crack depth have been discussed; the deep crack with large and small unbalance mass and the shallow crack with large unbalance mass. Finally, the vibration responses in the centroid of the shaft obtained from MBS have been exported into FE software to observe the breathing mechanism. The principle of the simulation using an integrated FE and MBS is summarized and presented in Figure 6.2.

6.1 Finite element model of flexible cracked shaft

The transverse crack model (for circular cross-section shaft) uses cohesive elements and is modeled by various relative crack depths for $a/d=$ 0.1 to 0.5. Length and diameter of shaft are 1.0 m and 0.08 m, respectively. About 5 000 continuum elements and 100 cohesive elements have been used. The cohesive elements along the crack surfaces are implemented as interface elements as shown in Figure 6.3.

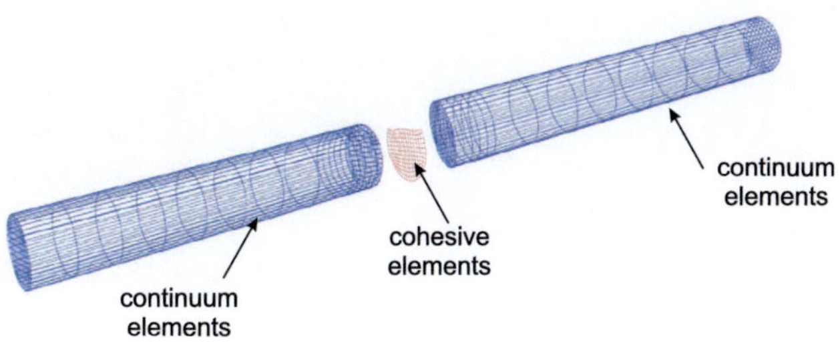

Figure 6.3: Finite element model of a flexible cracked shaft

The natural frequencies and mode shapes of a cracked shaft at rest (rotating speed $\Omega=0$) have been investigated. The methodology is to vary the relative crack depth from un-cracked shaft ($a = 0$) until the deepest relative crack depth ($a/d = 0.5$) in discrete steps. At each step (or, value of the relative crack depth), the normalised natural frequencies or ratio between natural frequency of the cracked shaft and natural frequency of the un-cracked shaft can be determined directly. The variations of the first three normalised natural frequencies ω_{cri}/ω_{ni}, $i = 1, 2, 3$ with relative crack depth are shown in Figure 6.4. The change in the first normalised natural frequency with crack present is significant whilst the change in the second normalised natural frequency is quite small because the crack location is close to the anti-nodal point of the first mode. It can also be seen that for a given relative crack depth, the change in the third normalised natural frequency is moderate and monotonically decreases with the increment of the relative crack depth.

The corresponding mode shapes of the first three natural frequencies are shown in Figure 6.5. For shallow crack depth $a = 0.1d$, all of its mode shapes almost coincide with the mode shapes of uncracked shaft. At deep crack $a = 0.5d$, the only difference to the mode shapes of the uncracked shaft occurs for the third only. That means that the differences between the cracked and uncracked mode shapes are very small and very difficult to detect in prac-tice. These results have a very good agreement with the mode shape results (Figure 5.8) obtained by using FE of one dimensional continuum rotor model as discussed in Chapter 5.

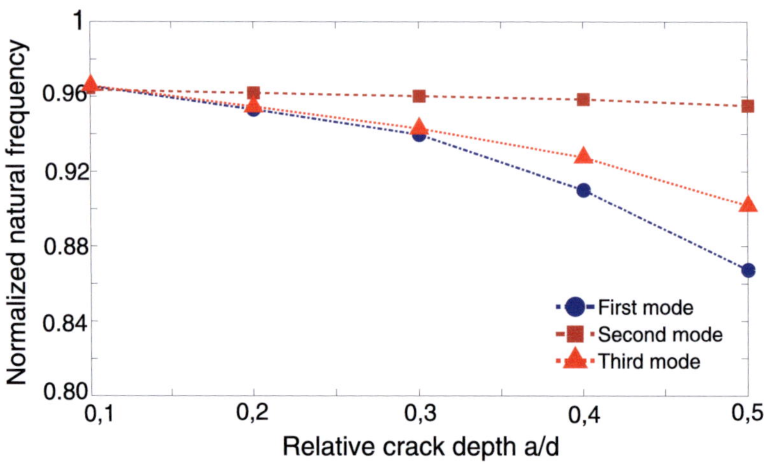

Figure 6.4: Normalised natural frequencies versus relative crack depth

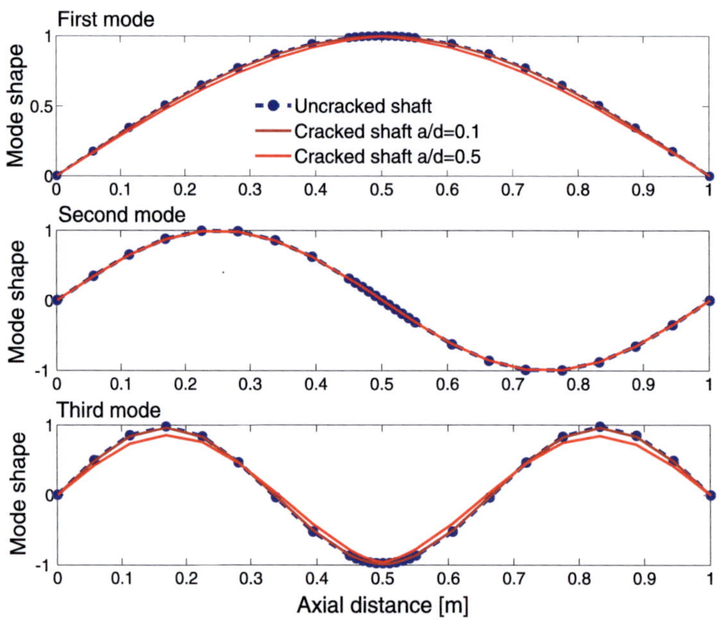

Figure 6.5: Comparison of the mode shapes between cracked and uncracked shaft

6.2 Dynamic behaviour of rotating flexible cracked shaft

The FE model of the elastic cracked shaft is exported into MBS in order to analyze the dynamic loads, due to the crack, unbalance and inertia force acting during rotation at different rotating speeds. The elastic cracked shafts are supported by two rigid bearings as shown in Figure 6.6. The dynamic analyses consist of cracked shaft loaded by weight only and by weight and unbalance. The effect of orientation angle of the unbalance mass on the breathing crack behaviour has also been investigated. The case where an unbalance is located on the crack side, and the case where the unbalance is located on the opposite side of the crack have been investigated. Darpe et al. [31], Cheng et al. [21], Yamamoto and Ishida [156] reported that the breathing behaviour and the peak response are strongly influenced by the unbalance orientation angle relative to the crack direction. In this work, three cases of unbalance masses have been studied; the deep crack with large and small unbalance mass and the shallow crack with large unbalance mass.

The aim of the study is to analyse the additional deflection due to the breathing crack during one revolution of the shaft and to study the effect of different unbalance orientation angles. Using MBS, the dynamic effects can be observed and taken into account. In the next section, the transverse vibration response results are used as deflection input for the FE model in order to predict the breathing mechanism during rotation of shaft.

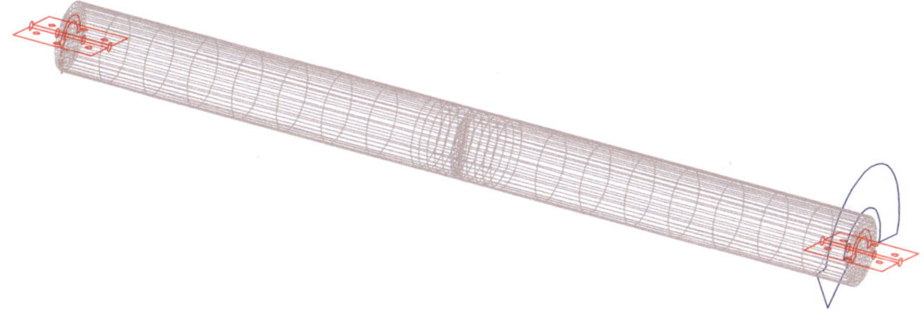

Figure 6.6: FE model of the elastic cracked shaft supported by rigid bearings in MBS

6.2.1 Flexible cracked shaft loaded by weight only

The deflection line of a shaft with a crack in tension zone is given by superposition of two parts: the deflection line of the uncracked shaft and the additional deflection caused by the local compliance of the crack as shown in Figure 6.7 [42]. This additional part cannot be found from the elastic beam theory, because a crack is the weakening of the bending stiffness on a zero length.

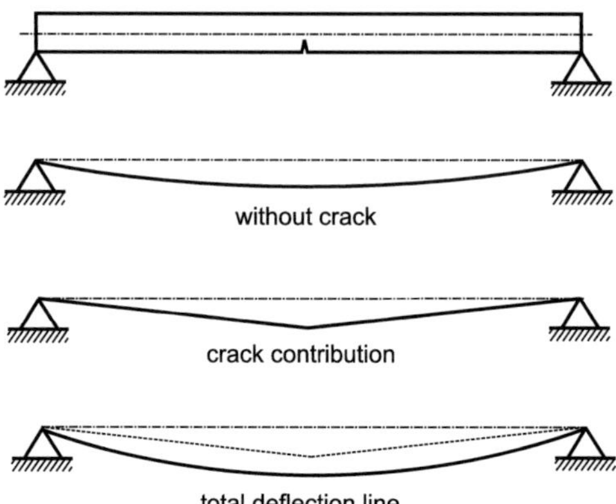

Figure 6.7: Deflection line due to contributions of the uncracked shaft and the local crack compliance

The breathing mechanism is the result of the stress and strain distribution around the cracked area [12], which is due to:

1. static loads like weight,

2. dynamic loads, due to the inertia force and the unbalance force.

The present analysis investigates the breathing modelled by means of FE (Figure 6.8) with the relative crack depth a/d=0.1 (shallow crack) and $a/d = 0.5$ (deep crack) at midspan between two rigid bearings.

The methodology is to vary the rotating speed from low speed to a very high speed in discrete steps using MBS software MSc.Adams. At each value of the rotating speed, the amplitudes in both lateral directions are observed. Figure 6.9 shows the vibration amplitude of a cracked shaft due to weight in lateral-vertical direction. As can be seen from Figure 6.9, by comparing the breathing crack with relative crack depth a/d=0.1 (shallow crack) to a/d=0.5 (deep crack), amplitude of shallow cracked shaft increases slowly by increasing rotating speed. The main difference with respect to the deep crack is that the increasing of amplitude in case of deep crack is non-linear and breathing behaviour is strongly influenced by the vibration and is not governed by weight. This may be due to the fact that at high rotating speed, the crack does not close completely once per revolution.

As the rotation speed increases, the vibration amplitude also increases whilst the static deflection remains constant which can be depicted by steady state orbital response at various speeds in Figure 6.10.

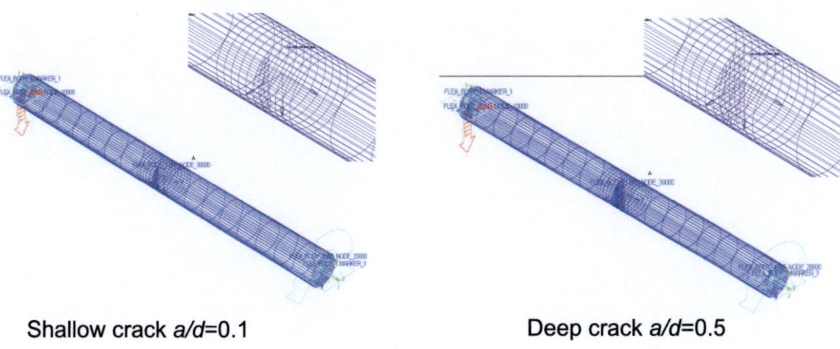

Figure 6.8: FE model of elastic cracked shafts loaded by weight only

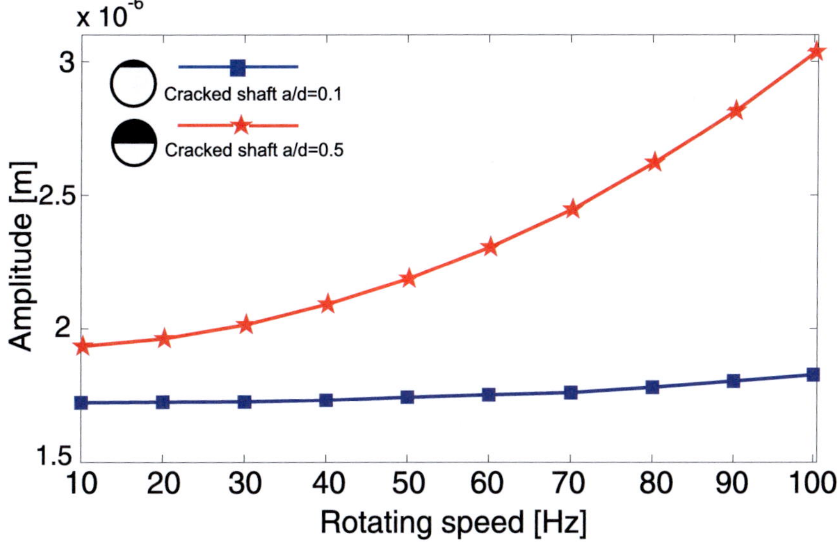

Figure 6.9: Vibration amplitude of flexible cracked shaft loaded by weight only at various
rotating speeds

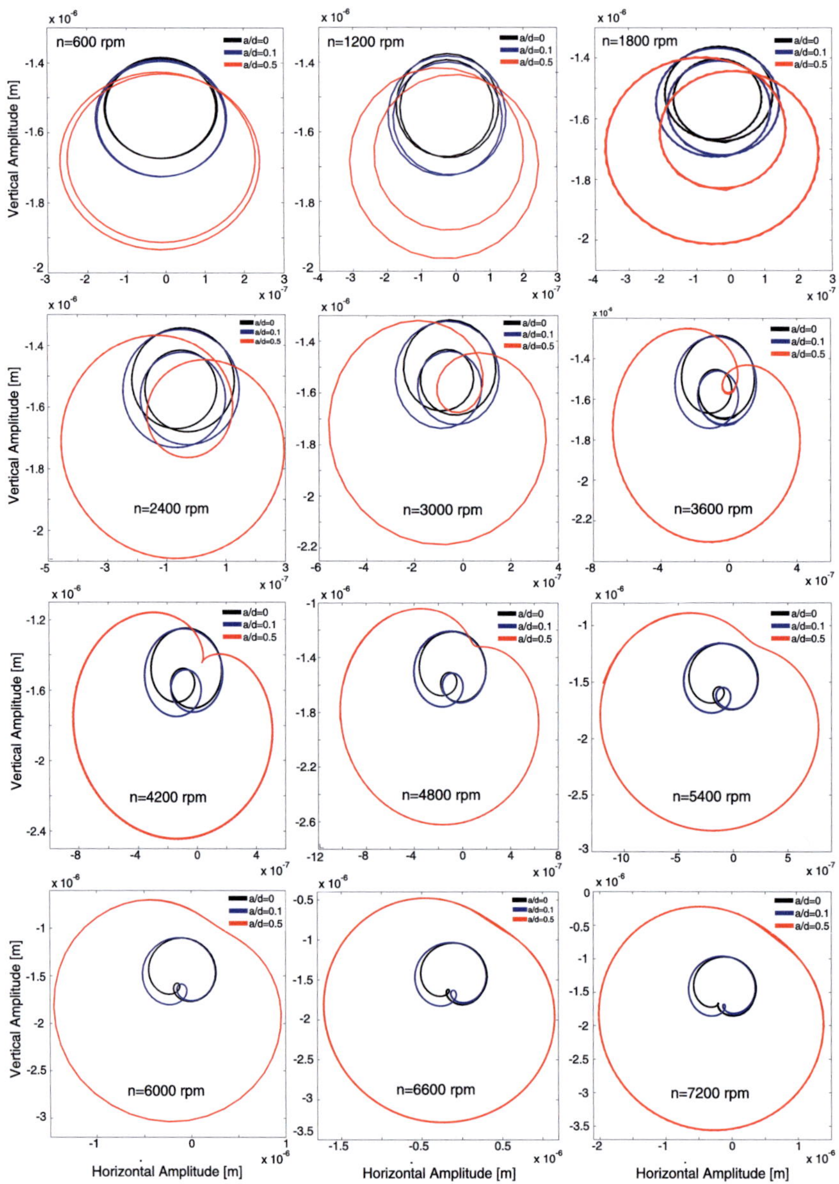

Figure 6.10: Steady state orbital responses of flexible cracked shaft loaded by weight only at various rotating speeds

6.2.2 Flexible cracked shaft loaded by weight and unbalance

Aim of this sub-section is to study not only the effect of weight but also the effect of unbalance, especially the effect of different unbalance orientations in case of a deep and a shallow crack with large and small mass unbalance. Two extreme cases of unbalance orientation are discussed. One is the case where an unbalance is located on the crack side, and the other is the case where this unbalance is located on the opposite side of the crack.

6.2.2.1 Case 1: Deep crack $a/d = 0.5$, large unbalance mass

The mass of the shaft and disk are selected to be 39.287 kg and 3.9287 kg, respectively. The unbalance is modelled by large unbalance mass, i.e. 1 kg and is mounted on the shaft at midspan between the two rigid bearings. As a first case, a symmetric elastic cracked shaft with deep crack $(a/d = 0.5)$ is studied. The model of the elastic cracked shaft and its geometry is displayed in Figures 6.11 and 6.12, respectively. For comparison, a symmetrical shaft without crack is also investigated.

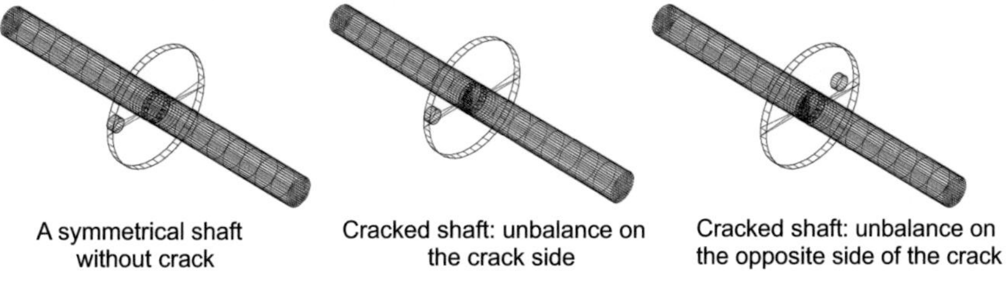

Figure 6.11: FE model of elastic uncracked and cracked shaft with unbalance in MBS

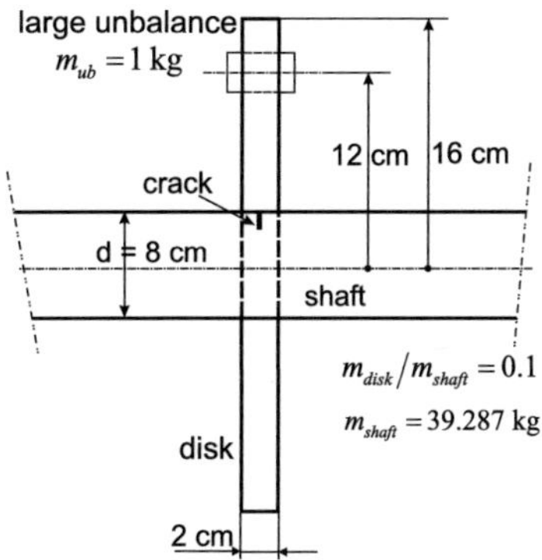

Figure 6.12: Geometry of cracked shaft model for relative crack depth a/d=0.5 with large unbalance mass

Vibration amplitude as function of rotating speed is shown in Figure 6.13. In case of large unbalance mass, the vibration amplitude of cracked rotor changes significantly, depending on the direction of the unbalance. If an unbalance is located on the same side as the crack, the vibration amplitude increases stronger than the vibration amplitude of the uncracked rotor. In contrast, if the unbalance is on the opposite side of the crack, the vibration amplitude is lower than the vibration amplitude of the uncracked rotor. This may be caused by the fact that the vibration amplitude due to the crack is opposite in direction to the vibration amplitude due to the unbalance force.

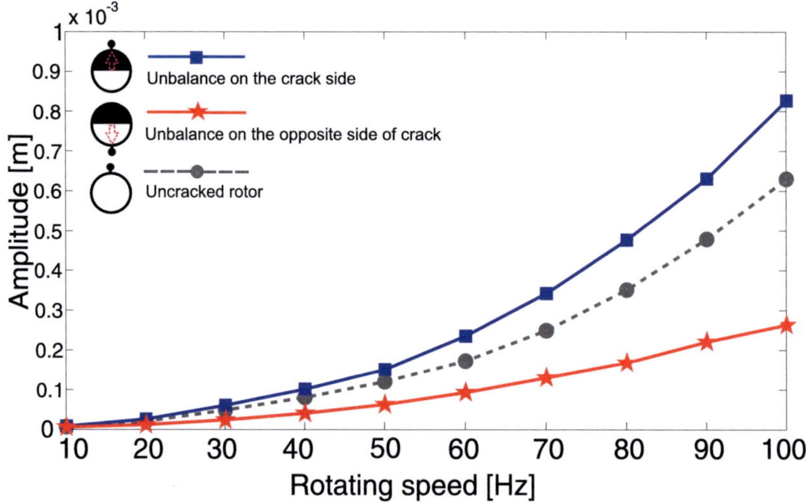

Figure 6.13: Vibration amplitude of flexible uncracked and cracked shaft loaded by weight and large unbalance at various rotating speeds: relative crack depth a/d=0.5

6.2.2.2 Case 2: Deep crack $a/d = 0.5$, small unbalance mass

Mass of shaft and disk are the same as in the previous simulation ($m_{disk}/m_{shaft} = 0.1$), but now the unbalance is modelled by a small unbalance mass, i.e. 0.1 kg. In this case, a symmetric elastic cracked shaft with deep crack ($a/d = 0.5$) is used. As comparison, a symmetrical shaft without crack is also investigated.

Figure 6.14 depicts the vibration amplitude as a function of the rotating speed. In case of a small unbalance mass, the vibration amplitude of the cracked rotor changes also with the direction of the unbalance. If the unbalance is located on the same side as the crack, the vibration amplitude increases significantly stronger than the vibration amplitude of the uncracked rotor. In case of an unbalance on the opposite side of the crack, the vibration amplitude increases also stronger than that of an uncracked rotor. This is because the breathing mechanism is governed by the weight than by the unbalance force.

6.2.2.3 Case 3: Shallow crack $a/d = 0.1$, large unbalance mass

In the third case, as for the first case, the mass of the shaft and the disk are the same and the unbalance is modelled by a large unbalance mass, but here a symmetric elastic cracked shaft with shallow crack ($a/d = 0.1$) is presented. In case of a shallow crack with large unbalance mass, both an unbalance located on the same side as the crack and on the opposite side of the crack, the vibration amplitude of cracked rotor increases always a little bit larger with increasing rotating speed than the vibration amplitude of the uncracked rotor. The small or shallow crack plays a minor role and has nearly no effect on the vibration amplitude. Thus, in this case the breathing mechanism is governed by vibration due to unbalance force rather than by the crack. Figure 6.15 shows the vibration amplitude curve as function of rotating speed.

In summary, the main results of the MBS of the reduced FE system in these two sections are the following:

1. In case of a shaft without unbalance mass, the breathing mechanism for a shallow crack ($a/d = 0.1$) is strongly governed by weight. On contrary, for a deep crack ($a/d = 0.5$), the breathing mechanism is governed by vibration rather than by weight.

2. In case of a shaft with unbalance mass, the vibration amplitudes strongly depend on unbalance orientation with respect to the crack.

3. The analyses are based on a rotating shaft, where crack opening and closing occurs and the crack breathes governed by rotation and vibration due to the inertia force. Implementation of the breathing mechanism using every node at the crack surface during rotation is more complicated and leads to a considerable CPU time. The breathing mechanism is known when the open and closed parts of the crack have been identified for any angle rotation. To investigate the breathing mechanism during one revolution of the shaft, the FE model is therefore used again. This study will be discussed in the next section.

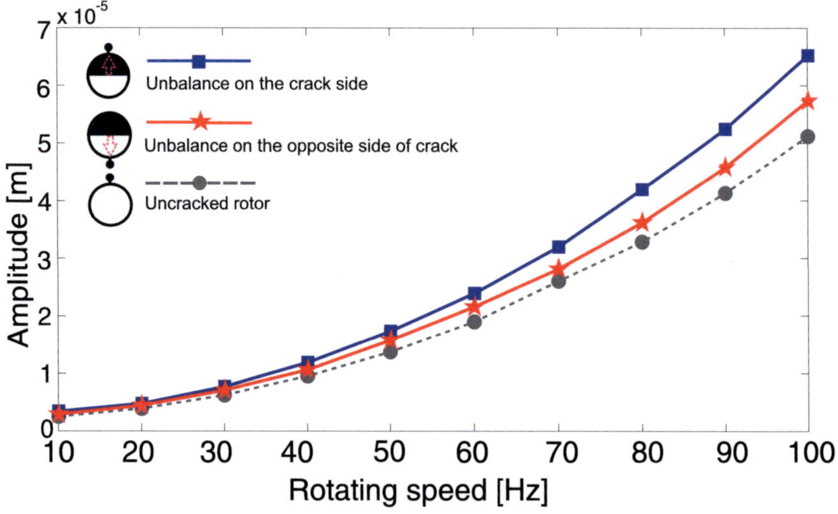

Figure 6.14: Vibration amplitude of flexible uncracked and cracked shaft loaded by weight and small unbalance at various rotating speeds: relative crack depth a/d=0.5

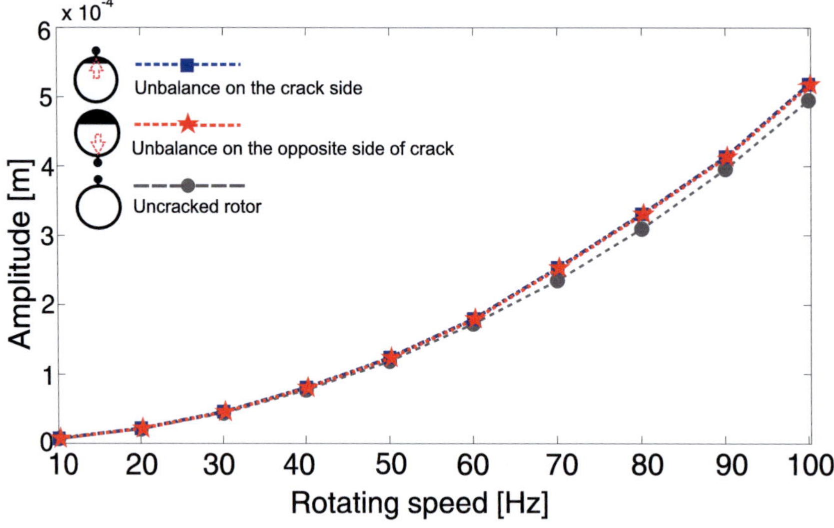

Figure 6.15: Vibration amplitude of flexible uncracked and shallow cracked shaft loaded by weight and large unbalance at various rotating speeds: relative crack depth a/d=0.1

6.3 Breathing mechanism in finite element simulation

The breathing mechanism can be predicted by means of 3D FE models by applying load conditions and evaluating deflections. The investigation requires a fine mesh to take into account the geometry of the crack, which will be time consuming. The 3D FE model allows obviously calculating deflections and strains, by taking into account the breathing mechanism. In this section, the investigation of the breathing crack mechanism is investigated using two methods, namely breathing crack under rotating loading (non rotating shaft) and breathing crack under rotating shaft (under inertia force).

6.3.1 Breathing crack under rotating loading

Transverse cracked shafts with relative crack depth $a/d = 0.1$ and 0.2 are considered, length and diameter of shaft are 1.0 m and 0.08 m, respectively. The breathing mechanism is generated by the bending due to external load (weight) by increasing the angle by steps of 15° ($\pi/12$ rad), i.e. 24 divisions for one revolution (i.e 24 steps in FE programm). The observation of opening crack is repeated for all different angular positions of the cracked shaft specimen. The breathing (open and closed crack areas are evaluated in each angular step) is observed by the nodal displacement and the stress distribution (tensile or compressive stress) around the crack.

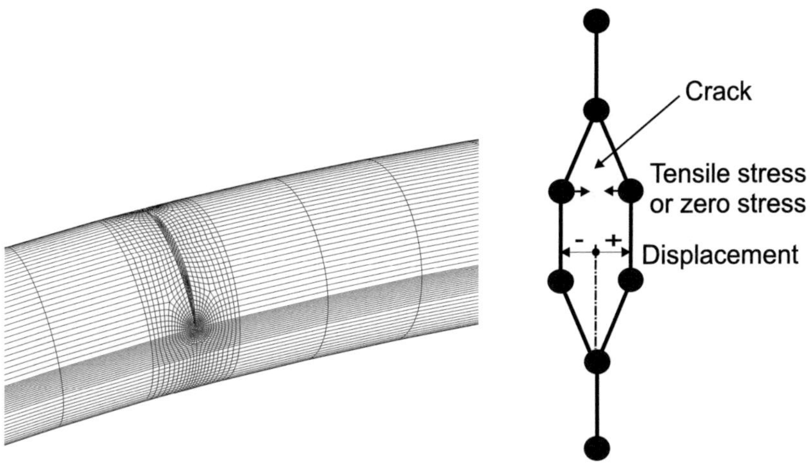

Figure 6.16: Evaluation of displacement and stress criteria for state of crack

The prediction of the breathing mechanism was performed according following steps:

- Rotating load due to heavy weight shaft is realized by increasing the angle by steps of 15° ($\pi/12$ rad).

- Stress distribution due to bending moment is known over the cross section.

- Closing crack is defined that due to compressive stresses, contact forces appear and the crack area element is closed. Crack opens while zero or very small numerical value of stresses appear at which there is no contact force.

- Displacements and stresses can be observed around crack surfaces.

- In order to avoid local deformations due to the application of loads, the deflection and stress distribution at each of the element crack area muss be evaluated as shown in Figure 6.16. That means two nodes defining crack have not only the same value of displacement with opposite direction but also have small positive or zero stress. If these two conditions at the node occur, it means the crack is open.

Figures 6.17 and 6.18 represent some results obtained during one cycle rotating load for a non-rotating shaft with relative crack depth $a/d = 0.1$ and 0.2, respectively. As can be seen the crack opens more slowly at the beginning, but increases its opening at $\pi/3$ rad (60°). At $5\pi/6$ rad (150°) it is already completely open. The crack closes again at $4\pi/3$ rad (240°) and increases its closing at $3\pi/2$ rad (270°). The crack is already completely closed at $11\pi/6$ rad (330°). As can be seen there are some relevant differences with respect to the proposed crack model discussed in Chapter 3. It can be noted that this method is not accurate because the observation has not taken into account the influence of whirl during shaft rotation.

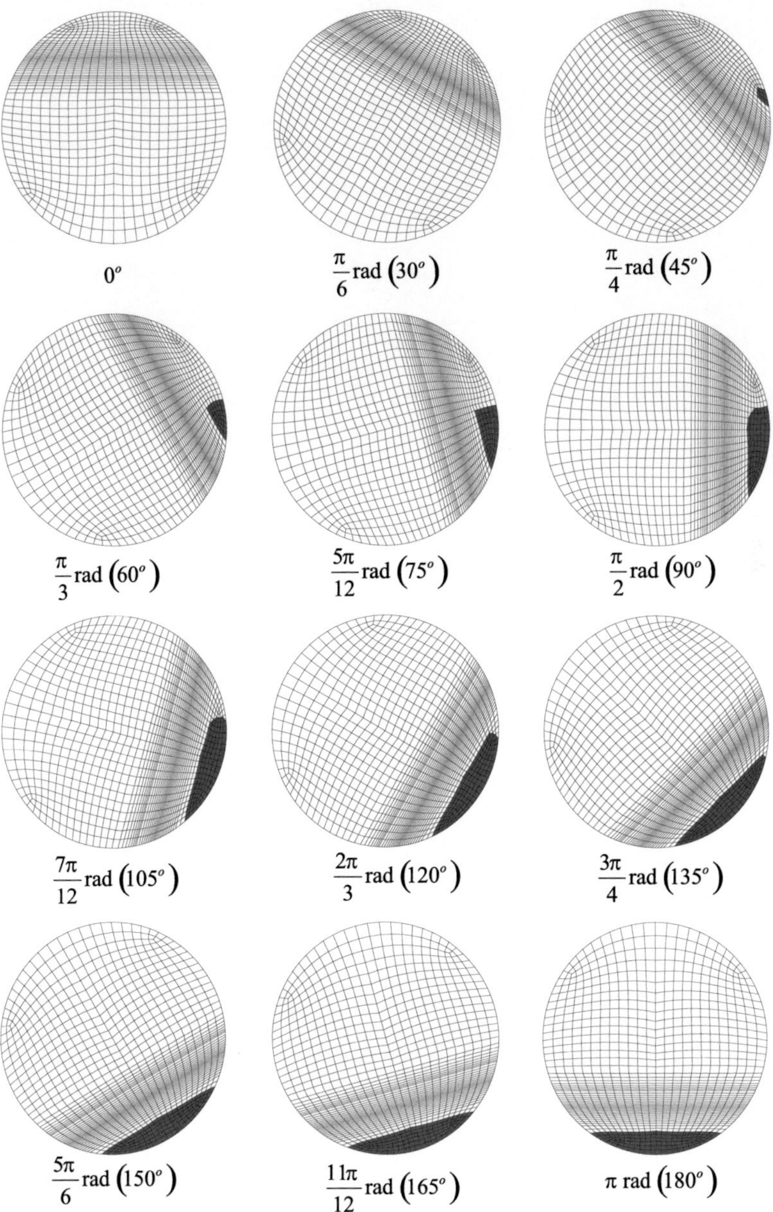

Figure 6.17: Breathing crack of the non rotating cracked shaft under rotating load for relative crack depth a/d=0.10

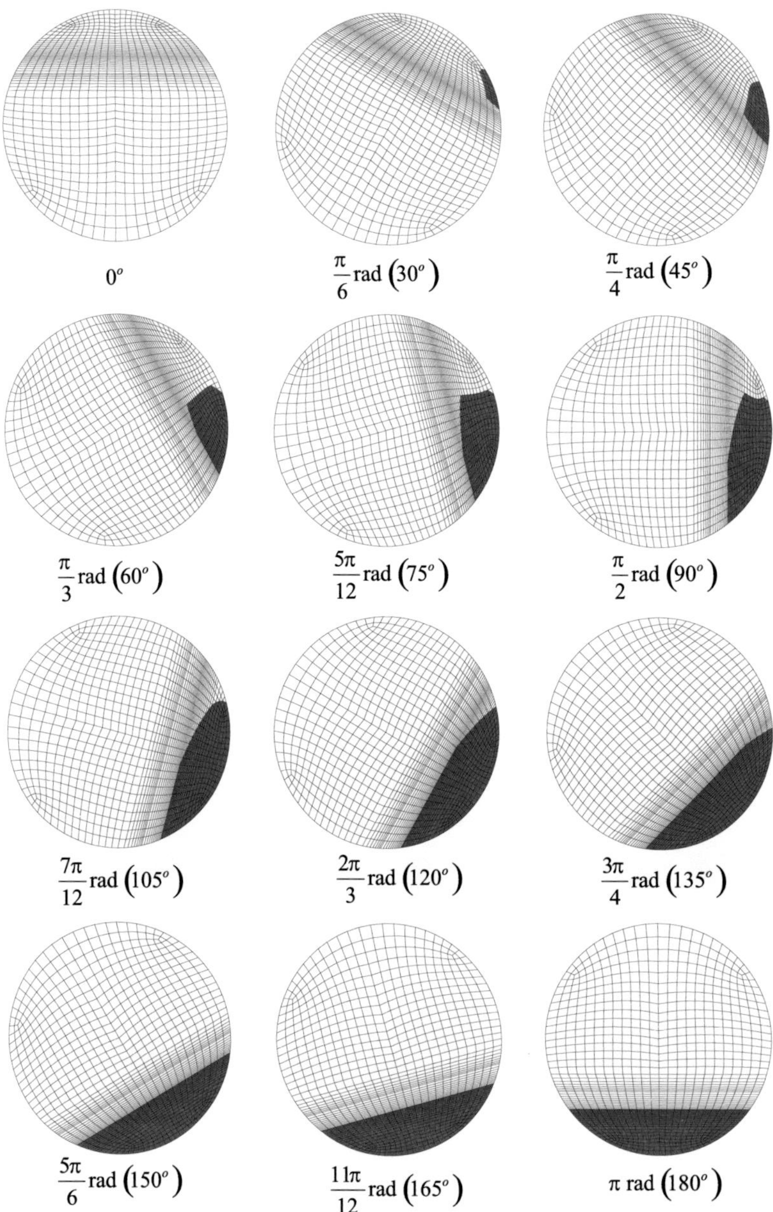

Figure 6.18: Breathing crack of the non rotating cracked shaft under rotating load for relative crack depth $a/d{=}0.20$

6.3.2 Breathing crack during rotation of the shaft

The breathing mechanism generated by the rotating bending load discussed in previous section has some limitations. Due to the presence of inertia forces, the dynamic behaviour of rotating structures is different from those of static structures. Although in case of weight dominance, the amplitude vibration response due to inertia force is smaller than due to weight forces, elastic forces and presence of a crack (shaft is assumed to be balanced). Computating the inertia force into account will yield more accurate results. The idea is that the vibration responses in the centroid of the shaft obtained from MBS software are exported into the FE software in order to analyse the breathing mechanism, as schematically shown in Figure 6.19. The opening crack is simulated for one cycle of revolution of the cracked shaft specimen in steady state condition. The breathing mechanism is generated using same technique as in the case of cracked shaft under rotating load.

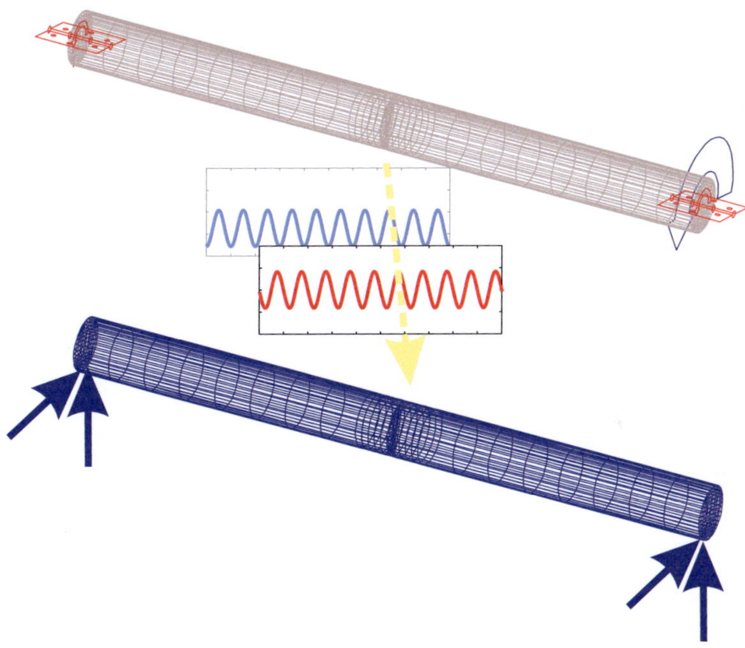

Figure 6.19: Export the vibration responses from MBS into FE

Both lateral vibration amplitude and steady state orbital responses of cracked shaft in case of relative crack depth $a/d=0.1$ at rotating speed 600 rpm (10 Hz) are displayed in Figures 6.20 and 6.21, respectively. Both amplitudes in lateral direction are employed in FE model by using step increment during one revolution of the rotor.

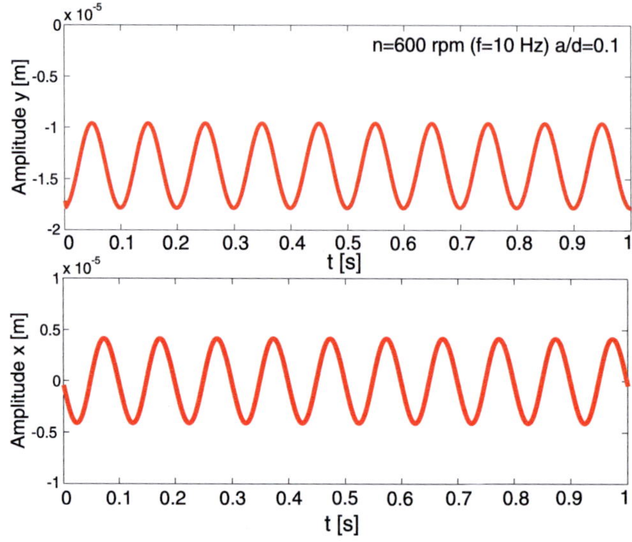

Figure 6.20: MBS result: Lateral vibration responses of a cracked rotor during rotation

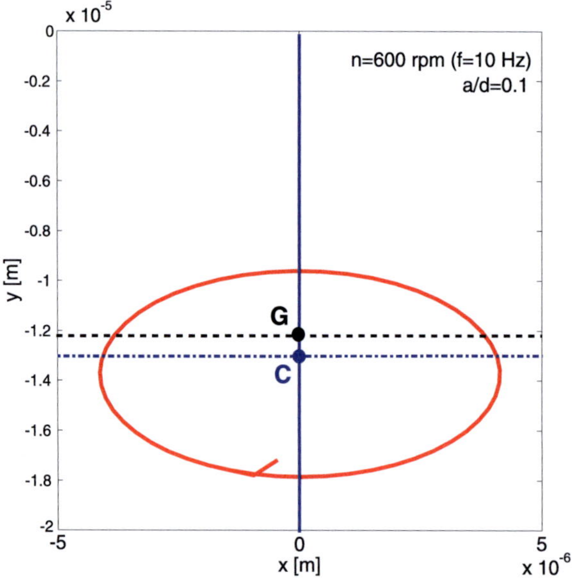

Figure 6.21: MBS result: Steady state orbital responses of a cracked rotor during rotation

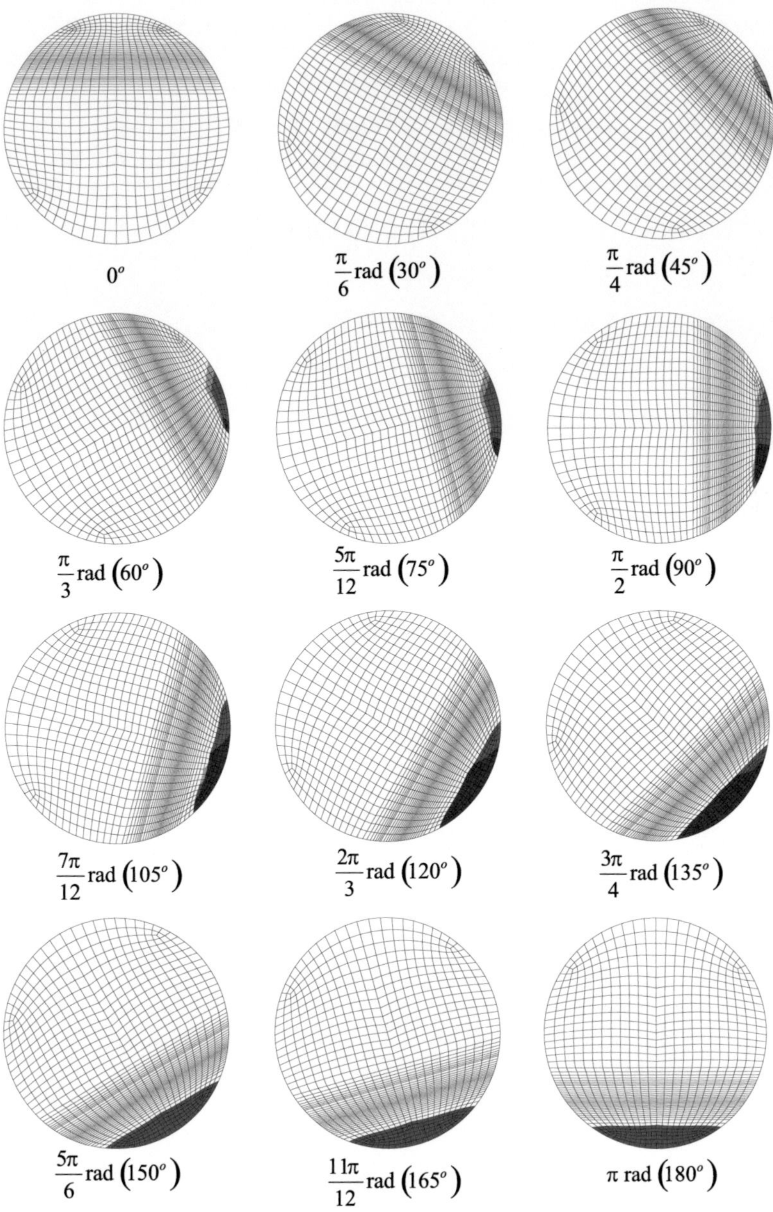

Figure 6.22: Breathing crack of the rotating cracked shaft for relative crack depth $a/d{=}0.10$

Table 6.1: Comparison simulation results of breathing mechanism between rotating shaft and rotating load for relative crack depth a/d=0.1 at rotating speed Ω=600 rpm

Characteristic	Case rotating shaft	Case rotating load
Crack will start to appear and begin to open	$\pi/6$ rad (30°)	$\pi/4$ rad (45°)
Crack depth opening	different stages of crack depth	direct to amount of crack depth
Direction crack opening	inclined to the crack front	perpendicular to the crack front
Velocity of crack opening	quite constant its opening crack area	increase its opening crack area at $\pi/3$
Completely open	$3\pi/4$ rad (135°)	$5\pi/6$ rad (150°)
Crack begins to close	$5\pi/4$ rad (225°)	$4\pi/3$ rad (240°)
Completely closed	$23\pi/12$ rad (345°)	$11\pi/6$ rad (330°)

Figure 6.22 displays the resulting breathing mechanism during half revolution of a rotating shaft (from closed crack to open crack). It can be observed that the simulation results of breathing mechanism under rotating shaft (Figure 6.22) and under rotating load (Figure 6.17) demonstrates some relevant differences and are summarized in Table 6.1. Figure 6.23 shows the simulation results under the influence of weight when the cracked shaft is rotated and the shaft whirls are taken into account. O is the bearing centerline, G is the shaft geometry center, $\overline{OG}$ is static deflection of the shaft due to its weight and $\overline{OC}$ is static deflection of the shaft due to its weight and due to presence of the crack. The shaft rotates along the red orbit line. The differences of the simulation results of breathing mechanism between rotating shaft and rotating load are caused by inertia force acting during shaft rotation.

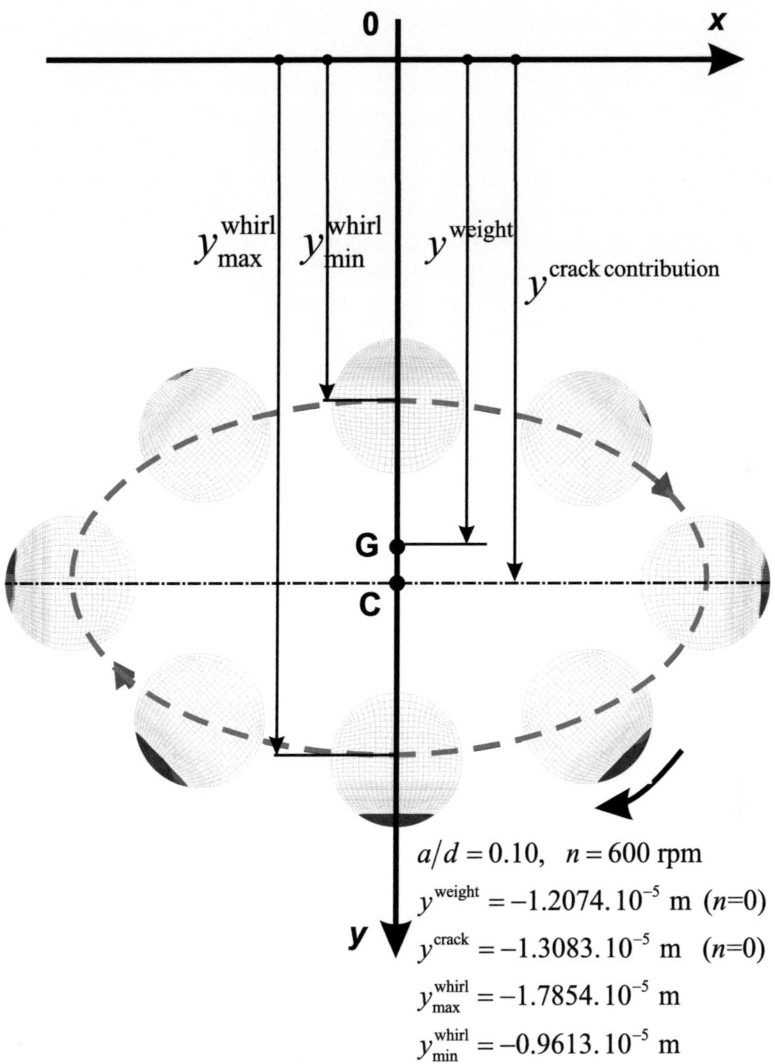

Figure 6.23: Rotating cracked shaft during one revolution

6.4 Validation of the breathing crack model

In comparison with the crack closure straight line model in Chapter 3, the simulation results are only different for rotation angles less than $\pi/2$ rad (90°). Figure 6.24 displays the comparison between the simulation results and the crack closure straight line model for shallow crack a/d=0.1, from the beginning until the crack opens completely. As discussed before in Chapter 3, crack opening of the crack closure straight line model increases with the same amount for every step of angle of rotation, i.e. $a(\theta) = a\theta/\pi$ where a is crack depth and θ angle of rotation until half revolution (π rad). It can be observed that crack opening of the straight line model increases faster than the simulation results until rotation angle $5\pi/12$ rad (75°). Crack opening of the straight line model is in good agreement with crack opening of the simulation results between angles $5\pi/12$ to $5\pi/6$ rad (75° to 150°). After that, crack opening of the simulation results opens more rapidly than the straight line model. These results were also obtained when crack closes.

It has been shown that the simulation results of breathing mechanism in rotating shafts can be accurately generated by integrated simulation process of FE models and MBS. The simulation results are based on transverse vibration responses of a rotating shaft from MBS, by reducing the number of degrees of freedom of a FE model. Nevertheless, the proposed model can be used for calculating the variation of the stiffness of a cracked shaft during one revolution accurately.

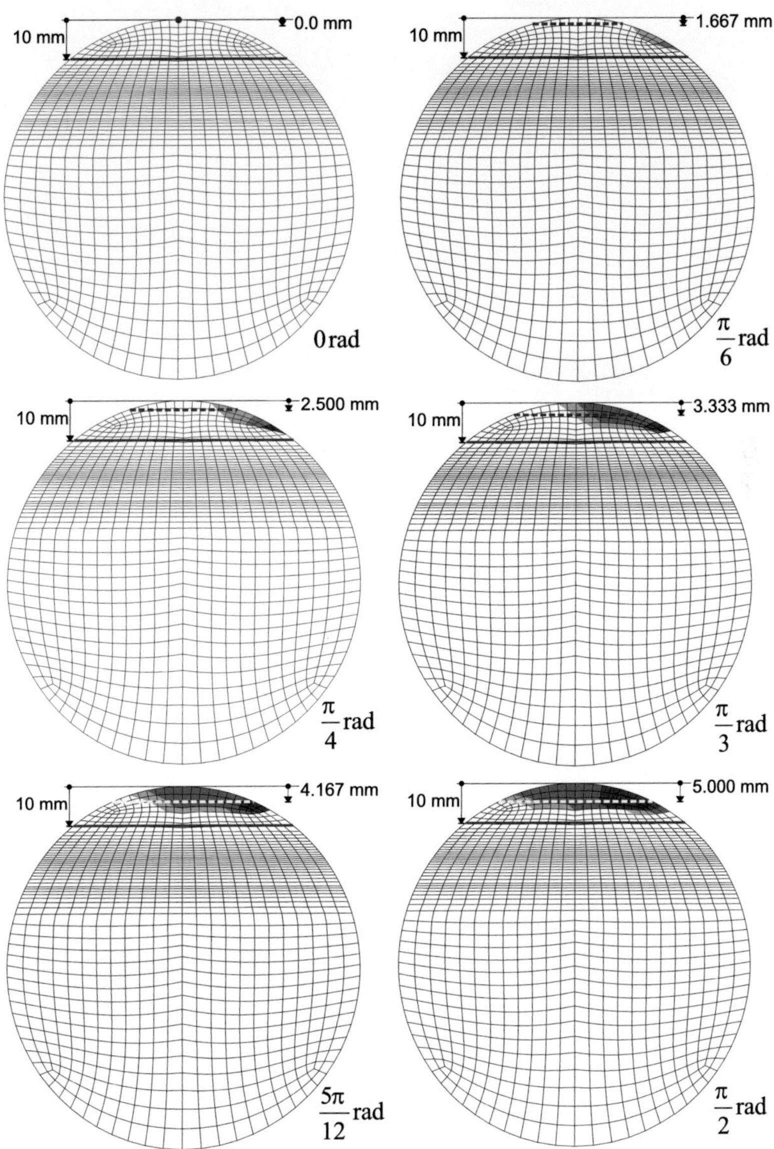

10 mm
0.0 mm
0 rad
10 mm
1.667 mm
π/6 rad
10 mm
2.500 mm
π/4 rad
10 mm
3.333 mm
π/3 rad
10 mm
4.167 mm
5π/12 rad
10 mm
5.000 mm
π/2 rad

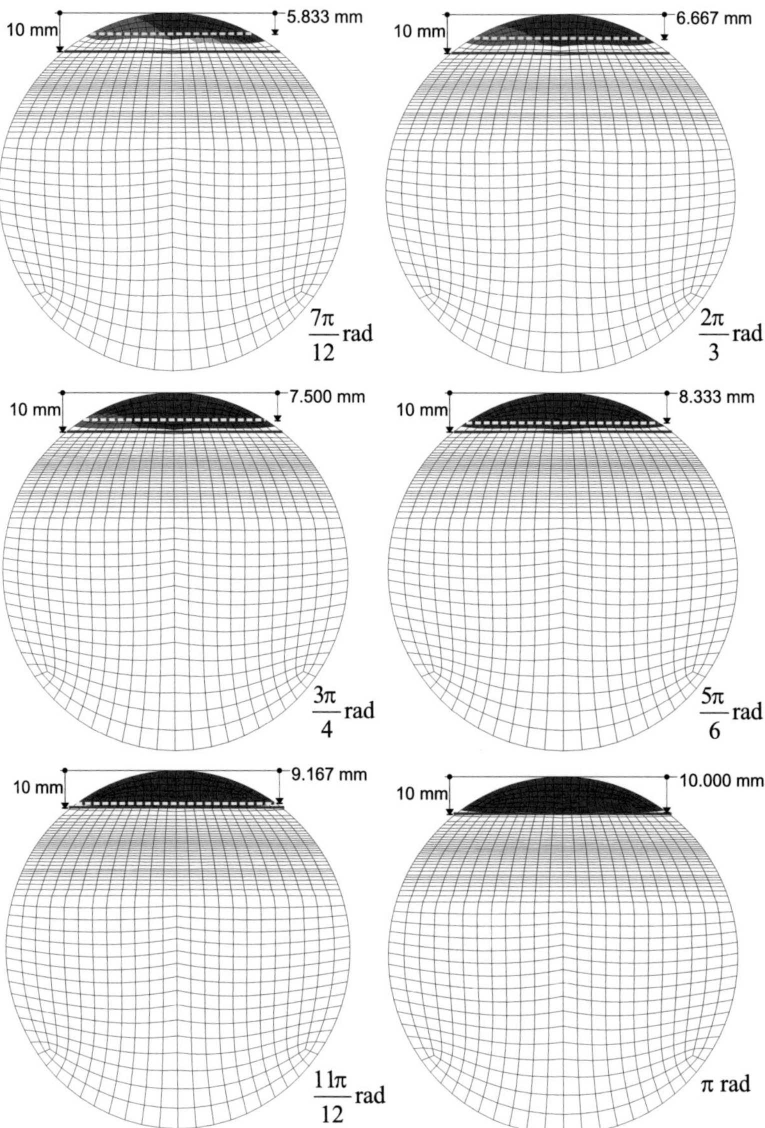

Figure 6.24: Comparison between the crack closure straight line model and the simulation results of breathing mechanism for relative crack depth $a/d{=}0.1$

7 Conclusions

7.1 Major results

The main objective of this work is the application of the cohesive zone model (CZM) to analyze and study the vibrational behaviour of a rotor with a transverse breathing crack. In this regard, four aspects have been investigated and proposed:

1. An explicit breathing steering function using CZM to estimate stiffness variation of cracked shaft during rotation.

2. Two breathing crack shapes, namely crack closure straight line and parabolic line are proposed for shallow crack and deep crack, respectively.

3. Two 1D FE models for cracked rotors based on CZM are proposed. The first model is based on an asymmetric matrix of area moment of inertia using a breathing steering function derived from a CZM, while the second one is based on zero thickness cohesive element between continuum elements.

4. An integrated simulation process of FE and MBS is developed to investigate the breathing mechanism of a cracked shaft.

In Chapter 2, the CZM is used to study the fracture process zone. This approach is based on energy balance and has the advantage that it can be implemented easily in any numerical procedure. The presented numerical applications demonstrated the efficiency of the CZM in simulating the crack problems properly. In the first case, one-cohesive element shows that the interface elements open when damage occurs and loose their stiffness at failure so that the continuum element is disconnected. In the second case, stress distribution on a cracked plate based on LEFM, Irwin's model, Dugdale's model and the CZM are compared. These simulations are made to show that the presence of cohesive elements in a zone ahead of the crack tip generates stresses that are lower than that in LEFM due to the softening of cohesive elements and due to the fact that the CZM can avoid stress singularity near crack tip. Furthermore, an implementation of 3D cohesive elements for delamination predictions coincides nearly with experiment results. The energy balance method, which was implemented for crack growth and crack propagation of a cracked plate, shows that the external work flows as recoverable elastic strain energy, inelastic strain energy, plastic dissipative energy and cohesive energy are the energy consuming mechanisms within the fracture process zone.

The main failure mechanism in ductile metals consists of the nucleation of voids and their growth and coalescence that initiates at the inclusions and second phase particles. Central to the growth of these voids is the triaxiality of the stress state. On the basis of a triaxiality dependent CZM, the changes of direct stiffnesses of the cracked shaft during rotation have also been investigated in Chapter 2. To determine the stiffness variations during one revolution of the cracked shaft, the transverse crack is assumed to be at the mid span of the shaft. The additional deflections in the rotating coordinates due to the crack are determined using Castigliano's theorem, where a versatile cohesive zone model to predict additional deflections at different states of stress is proposed. This model is developed for mode-I plane strain and accounts for triaxiality of the stress state explicitly by using CZM relations. For a particular angle of rotation, the CZM is introduced as a function of shaft geometry, shaft modulus of elasticity and yield strength, crack depth, lateral force and parameters of CZM. The proposed numerical solutions are compared to the breathing crack model based on LEFM. Since breathing crack modeled by LEFM has some limitations, the CZM gives more realistic results because it is based on cohesive energy which is valid as crack opens and closes.

In Chapter 3, based on FE models and the reported experimental results [11], [12], the breathing crack mechanism during rotation was modelled by a parabolic shape, which is the more realistic model. It was then shown that the parabolic breathing crack shape is considerably more general and accurate than the previously used functions in the literature. It can be noted that as long as the relative crack depth is small, the model of breathing crack parallel to crack front line or crack closure straight line may be used while the crack closure parabolic line should be used in case of deep crack.

During rotation of the rotor, a crack will open and close. As the crack opens, the shaft is locally asymmetric and this condition can lead to instability problems. The stability of a simple rotor system (de Laval rotor) due to a breathing crack has been investigated in Chapter 4. To focus on crack influence alone, crack-disk imbalance interaction and internal damping are ignored. In order to obtain the boundaries of stability regions the perturbation method has been applied. It is noticed that some small damping in the rotor system is very helpful to guarantee stability.

In Chapter 5, two FE model approaches of a cracked rotor based on equivalent beam using CZM have been proposed. Comparison with the results in literature show good agreement, as long as a single crack with regular shape, i.e. rectilinear shape is considered. Stiffness variation is induced by crack breathing which is function of the TSL. In the first proposed FE model, breathing is modelled by a function of the angular position that is called the breathing steering function, obtained by curve fitting. The second proposed FE model implemented one zero thickness element which is placed between continuum elements. The properties of zero thickness element is defined by the TSL. The crack closure straight line for shallow crack and parabolic line in case of deep crack are used which describes the cracked area during one revolution of the shaft. The second proposed FE model is more realistic than the first FE model since crack is the weakening of the bending stiffness on a length zero [42]. Therefore, using zero thickness cohesive element in FE model is reasonable and more realistic. Results obtained from CZM are compared with those obtained from the proposed model by Mayes and Davis [90], in which they used the cosine function. It seems that the CZM for cracks in a rotor is sufficiently accurate for health monitoring purposes and it can provide useful and robust information for crack identification.

The breathing mechanism of a simple cracked shaft on rigid supports has been studied in Chapter 6. An integrated simulation process of FE and MBS is employed. The idea is that an elastic cracked shaft with various relative crack depths is modelled by FE. Here, a breathing crack under rotating load (non-rotating shaft) is investigated. Then, the FE model of elastic cracked shafts is exported into MBS in order to analyze the dynamic loads, due to the crack, unbalance and inertia force acting during rotation at different rotating speeds. The effect of orientation angle of the unbalance mass on the breathing crack behaviour has also been investigated. Finally, the vibration responses in the centroid of the shaft obtained from MBS software have been exported again into FE software to observe the breathing mechanism. The main results of the 3D FE simulation in MBS software are the following:

Flexible cracked shaft loaded by weight and unbalance for a deep crack In case of a rotor without unbalance mass, the breathing mechanism for a shallow crack (a/d = 0.1)is strongly governed by weight. On contrary, for a deep crack (a/d = 0.5), the breathing mechanism is governed by vibration rather than by weight. In case of a rotor with a large unbalance mass, the vibration amplitudes strongly depend on unbalance orientation with respect to the crack. If an unbalance is located on the same side as the crack, the vibration amplitude increases stronger than the vibration amplitude of the uncracked rotor. In contrast, if the unbalance is on the opposite side of the crack, the vibration amplitude is lower than the vibration amplitude of the uncracked rotor. This may be caused by the fact that the vibration amplitude due to the crack is opposite in direction to the vibration amplitude due to the unbalance force. In case of a rotor with a small unbalance mass, the vibration amplitude of the cracked rotor changes also with the direction of the unbalance. If the unbalance is located on the same side as the crack, the vibration amplitude increases significantly stronger than the vibration amplitude of the uncracked rotor, while for the unbalance on the opposite side of the crack, the vibration amplitude increases also stronger than that of an uncracked rotor. This is because the breathing mechanism is governed rather by weight than by unbalance force.

Flexible cracked shaft loaded by weight and unbalance for a shallow crack In case of a shallow crack with large unbalance mass, on the same side as the crack or on the opposite side of the crack, the vibration amplitude of the cracked rotor increases always a little bit larger with increasing rotating speed than the vibration amplitude of the uncracked rotor. The small or shallow crack plays a minor role and has nearly no effect on the vibration amplitude. Thus, in this case the breathing mechanism is governed by vibration due to unbalance force rather than by the crack.

Breathing mechanism in FE-MBS simulation The simulation results have shown that the breathing mechanism is influenced by the vibration due to inertia forces, by rotating speed and by relative crack depth. It is shown that the relative crack depth during crack opening should be understood not as linear increasing. It can be noted that as long as the relative crack depth is small, the model of breathing crack parallel to crack front line (crack closure straight line model) may be used. The main difference with respect to the crack closure straight line is that the opening crack in the simulation results is not constant at the beginning. These results are also similar obtained when crack closes. The simulation results is in good agreement with the crack closure straight line model between rotation angles $\pi/3$ rad (60°) and $5\pi/6$ rad (150°).

7.2 Recommendations and further analysis

In light of the presented results and the conclusions, the following ideas can be recommended for future research that may extend the research presented in this work.

- CZM could be extended to consider plasticity and crack propagation of the cracked shaft. One important point on which the knowledge could be improved is the prediction of crack propagation on cracked rotor and residual life estimation from static loads and from the dynamical behaviour of cracked rotors.

- CZM has the advantage of its easiness of implementation in FE model to analyze the dynamic behaviour of a cracked shaft. It is recommended to use CZM to study different types of cracks such as longitudinal and slant cracks.

- Some other parameters such as internal damping, unbalances and thermal transients, could be studied to obtain results for their effect on breathing mechanism as well as stability of vibration.

- Further analysis on crack morphology is extremely important to understand the dynamic behaviour of cracked shaft. This would include shallow and wide cracks.

- Effect of rotating speed on breathing mechanism could be of interest.

A Analytical methods for rotating shafts with open crack

A.1 Timoshenko beam theory for rotating shaft

There are some different assumptions corresponding to theories of transverse vibrations of beams in literature [69], [149]. The most popular of fundamental theories is the Euler-Bernoulli beam-theory (1744). This theory takes into account the inertia forces due to the transverse translation and neglects the effect of shear and rotary inertia. Furthermore, this elementary beam theory is valid only when the height of the beam is small compared with its length. Another one is Rayleigh beam-theory (1877) which takes into account the effect of rotary inertia. Bress beam-theory (1859) and Volterra beam-theory (1955) take into account the rotational inertia, shear deformation and their combined effect. Difference between these theories, the bending stiffness of the beam according to the Volterra theory is $(1 - \nu^2)^{-1}$ times greater than that given by the Bress theory, where ν is Poisson ratio. This is because transverse compressive and tensile stresses are not allowed in the Volterra theory. Ambartsumyan beam-theory (1956) allows the distortion of the cross-section. Timoshenko beam-theory (1953) takes into account the rotational inertia, shear deformation and their combined effects. The fundamental difference between the Rayleigh and Bress theories, on one hand, and the Timoshenko theory, on the other, is that the correction factor in the Rayleigh and Bress theories appears as a result of shear and rotary effects, whereas in the Timoshenko theory, the correction factor is introduced in the initial equations.

An approach for bending vibration based on the Euler-Bernoulli beam theory has limitations because the analysis does not include the rotary inertia and shear deformation of the cross section. The effect of the rotary inertia and shear deformation reduces the fundamental natural frequency by 0.3% in a uniform beam with a radius-to-length ratio of 1:20, and the effect is bigger for higher modes [65]. The larger the radius-to-length ratio, the bigger the effect of the rotary inertia and shear deformation is on the fundamental natural frequency. Thus, the Timoshenko beam theory is applied to a general rotating shaft for an accurate analysis. The explicit natural frequencies of an open cracked shaft based on fracture mechanics can be calculated by Timoshenko beam theory (Chasalevris and Papadopoulos [19], Jun and Gadala [64], Jun and Kim [65] and Tsai and Wang [141]). Wauer [148] studied the dynamic of a cracked, distributed parameter rotor component by using a rotating Timoshenko shaft. He proposed an analytical approach to generate model equations which can be used as a subset within an extensive system of equations of motion for a complex multi-shaft, multi-bearing rotating machine.

157

The equations of motion of a rotating uniform shaft based on Timoshenko beam theory, including the effect of the rotary inertia, transverse shear, gyroscopic moments and axial or tangential follower torque are [76], [65].

$$
EI\frac{\partial^4 y}{\partial z^4} - \frac{EI\rho}{\kappa G}\frac{\partial^4 y}{\partial z^2 \partial t^2} + T\frac{\partial^3 x}{\partial z^3} - \frac{T\rho}{\kappa G}\frac{\partial^3 x}{\partial z \partial t^2} + \rho A\frac{\partial^2 y}{\partial t^2}
$$
$$
- \rho A r_0^2 \left[\left(\frac{\partial^4 y}{\partial z^2 \partial t^2} - \frac{\rho}{\kappa G}\frac{\partial^4 y}{\partial t^4} \right) + 2\Omega \left(\frac{\partial^3 x}{\partial z^2 \partial t} - \frac{\rho}{\kappa G}\frac{\partial^3 x}{\partial t^3} \right) \right] \tag{A.1}
$$

$$
EI\frac{\partial^4 x}{\partial z^4} - \frac{EI\rho}{\kappa G}\frac{\partial^4 x}{\partial z^2 \partial t^2} - T\frac{\partial^3 y}{\partial z^3} + \frac{T\rho}{\kappa G}\frac{\partial^3 y}{\partial z \partial t^2} + \rho A\frac{\partial^2 x}{\partial t^2}
$$
$$
- \rho A r_0^2 \left[\left(\frac{\partial^4 x}{\partial z^2 \partial t^2} - \frac{\rho}{\kappa G}\frac{\partial^4 x}{\partial t^4} \right) - 2\Omega \left(\frac{\partial^3 y}{\partial z^2 \partial t} - \frac{\rho}{\kappa G}\frac{\partial^3 y}{\partial t^3} \right) \right] \tag{A.2}
$$

where x and y are the displacements in the horizontal and vertical directions and z the axial coordinate, respectively (Figure 5.1). E, G, and ρ are modulus of elasticity, shear modulus and mass density, respectively. A and I are the area and area moment of inertia of the cross-section, r_0 the radius of gyration, κ the form factor or shear coefficient, T the torque on each end of the shaft and Ω the rotating speed.

It is noted that κ is the shear coefficient occurring in Timoshenko's differential equation for flexural vibrations of beams. Many authors [71], [138], [115], [140] used the shear factor for Timoshenko's beam with circular cross section

$$
\kappa = \frac{6\,(1+\nu)}{7+6\nu} \tag{A.3}
$$

Kaneko [67], [68] reported that expressions of κ for circular cross section are tabulated. The validity of the shear coefficient has been experimentally tested using a large number of cylindrical beams. He also presented that his experimental result seems to assure constancy of the shear coefficient over a fairly wide frequency range. It is pointed out that the expression of κ for a circular cross section is

$$
\kappa = \frac{6 + 12\nu + 6\nu^2}{7 + 12\nu + 4\nu^2} \tag{A.4}
$$

where ν is Poisson's ratio. For $\nu = 0.3$ yields $\kappa = 0.925$.

Since the equations are coupled, the solution of y and x cannot be obtained simply. To avoid solving the coupled equations the dynamic behaviour in the y-z and x-z planes can be expressed by one equation in terms of the following complex variable

$$
u\,(z,\,t) = y\,(z,\,t) + jx\,(z,\,t) \tag{A.5}
$$

where $y(z,t)$ and $x(z,t)$ are the vertical and horizontal of the axial coordinate z and time t. Substituting Eq.(A.5) into Eqs.(A.1) and (A.2) one obtains

$$EI\frac{\partial^4 u}{\partial z^4} - \frac{EI\rho}{\kappa G}\frac{\partial^4 u}{\partial z^2 \partial t^2} - jT\frac{\partial^3 u}{\partial z^3} + j\frac{T\rho}{\kappa G}\frac{\partial^3 u}{\partial z \partial t^2} + \rho A\frac{\partial^2 u}{\partial t^2}$$
$$-\rho A r_0^2\left[\left(\frac{\partial^4 u}{\partial z^2 \partial t^2} - \frac{\rho}{\kappa G}\frac{\partial^4 u}{\partial z^4}\right) - j2\Omega\left(\frac{\partial^3 u}{\partial z^2 \partial t} - \frac{\rho}{\kappa G}\frac{\partial^3 u}{\partial t^3}\right)\right] = 0 \qquad (A.6)$$

Reordered according to the order of the derivatives with respect to z

$$EI\frac{\partial^4 u}{\partial z^4} - jT\frac{\partial^3 u}{\partial z^3} - \left(\frac{EI\rho}{\kappa G} + \rho A r_0^2\right)\frac{\partial^4 u}{\partial z^2 \partial t^2} + j2\rho A r_0^2 \Omega\frac{\partial^3 u}{\partial z^2 \partial t}$$
$$+j\frac{T\rho}{\kappa G}\frac{\partial^3 u}{\partial z \partial t^2} + \frac{\rho^2 A r_0^2}{\kappa G}\frac{\partial^4 u}{\partial t^4} - j2\frac{\rho^2 A r_0^2 \Omega}{\kappa G}\frac{\partial^3 u}{\partial t^3} + \rho A\frac{\partial^2 u}{\partial t^2} = 0 \qquad (A.7)$$

which is a fourth order partial differential equation of $u(z,t)$

The boundary conditions are expressed in terms of the displacement, slope of the shaft, bending moment, and shearing force as follows

$$\text{displacement} \quad : \quad u(z,t) \qquad\qquad\qquad\qquad\qquad\qquad\qquad (A.8)$$

$$\text{slope} \quad : \quad \frac{\partial u}{\partial z} \qquad\qquad\qquad\qquad\qquad\qquad\qquad (A.9)$$

$$\text{bending moment} \quad : \quad EI\frac{\partial^2 u}{\partial z^2} - jT\frac{\partial u}{\partial z} \qquad\qquad\qquad\qquad (A.10)$$

$$\text{shearing force} \quad : \quad EI\frac{\partial^3 u}{\partial z^3} - jT\frac{\partial^2 u}{\partial z^2} - \rho A r_0^2\left(\frac{\partial^3 u}{\partial z \partial t^2} - 2j\Omega\frac{\partial^2 u}{\partial z \partial t}\right) \qquad (A.11)$$

The fourth order partial differential equation can be solved analytically by separation of variables. The time-dependent harmonic motion of natural frequency ω can be separated in the variable $u(z,t)$ as

$$u(z,t) = U(z)e^{j\omega t} \qquad\qquad\qquad\qquad\qquad\qquad (A.12)$$

Substituting Eq.(A.12) into Eq.(A.7) results in the following equation

$$EI\frac{\partial^4 U}{\partial z^4} - jT\frac{\partial^3 U}{\partial z^3} + \left[\left(\frac{EI\rho}{\kappa G} + \rho A r_0^2\right)\omega^2 - 2\rho A r_0^2 \Omega\omega\right]\frac{\partial^4 U}{\partial z^2}$$
$$-j\frac{T\rho}{\kappa G}\omega^2\frac{\partial U}{\partial z} + \left[\frac{\rho^2 A r_0^2}{\kappa G}\omega^4 - 2\frac{\rho^2 A r_0^2 \Omega}{\kappa G}\omega^3 - \rho A\omega^2\right]U = 0 \qquad (A.13)$$

which is a fourth order ordinary differential equation of a complex variable U and complex coefficients in the following form

$$\frac{\mathrm{d}^4 U}{\mathrm{d}z^4} + a_1\frac{\mathrm{d}^3 U}{\mathrm{d}z^3} + a_2\frac{\mathrm{d}^2 U}{\mathrm{d}z^2} + a_3\frac{\mathrm{d}U}{\mathrm{d}z} + a_4 U = 0 \qquad\qquad (A.14)$$

where

$$a_1 = -\frac{jT}{EI}$$

$$a_2 = \frac{1}{EI}\left[\left(\frac{EI\rho}{\kappa G} + \rho Ar_0^2\right)\omega^2 - 2\rho Ar_0^2\Omega\omega\right]$$

$$a_3 = -j\frac{1}{EI}\frac{T\rho}{\kappa G}\omega^2$$

$$a_4 = \frac{1}{EI}\left[\frac{\rho^2 Ar_0^2}{\kappa G}\omega^4 - 2\frac{\rho^2 Ar_0^2\Omega}{\kappa G}\omega^3 + \rho A\omega^2\right]$$

(A.15)

Eq.(A.14) has solutions of the form

$$U = pe^{\lambda z}$$

(A.16)

Substituting Eq.(A.16) into Eq.(A.14) yields the fourth order polynomials of λ, which is also complex

$$\lambda^4 + a_1\lambda^3 + a_2\lambda^2 + a_3\lambda + a_4 = 0$$

(A.17)

For $T = 0$, we have

$$\frac{\mathrm{d}^4 U}{\mathrm{d}z^4} + a_2\frac{\mathrm{d}^2 U}{\mathrm{d}z^2} + a_4 U = 0$$

(A.18)

with

$$k = \pm j\frac{1}{2}\left(a_2 \pm \sqrt{a_2^2 - 4a_4}\right)^{\frac{1}{2}}$$

(A.19)

Define

$$r_1 = \frac{1}{2}\left(a_2 - \sqrt{a_2^2 - 4a_4}\right)^{\frac{1}{2}}$$

(A.20)

$$r_2 = \frac{1}{2}\left(a_2 + \sqrt{a_2^2 - 4a_4}\right)^{\frac{1}{2}}$$

(A.21)

We obtain the solution

$$\begin{aligned}
U(z) &= p_1 e^{k_1 z} + p_2 e^{k_2 z} + p_3 e^{k_3 z} + p_4 e^{k_4 z}\\
&= p_1 e^{-jr_1 z} + p_2 e^{-jr_2 z} + p_3 e^{jr_1 z} + p_4 e^{jr_2 z}\\
&= C_1\cos r_1 z + C_2\cos r_2 z + jC_3\sin r_1 z + jC_4\sin r_4 z
\end{aligned}$$

(A.22)

For initial value simply supported

$$U(0) = U''(0) = 0$$

(A.23)

$$U(L) = U''(L) = 0$$

(A.24)

Substituting initial values Eqs.(A.23-24) into Eq.(A.22) and its second derivative, we have

$$\left(r_1^2 - r_2^2\right) \sin r_1 L = 0 \tag{A.25}$$

or

$$
\begin{aligned}
\sin r_1 L &= 0 \\
r_1 L &= n\pi \quad (n = 1, 2, \ldots) \\
\frac{1}{2}\left(a_2 - \sqrt{a_2^2 - 4a_4}\right)^{\frac{1}{2}} &= \frac{n\pi}{L} \\
a_4 - \left(\frac{n\pi}{L}\right)^2 a_2 + \left(\frac{n\pi}{L}\right)^4 &= 0
\end{aligned}
\tag{A.26}
$$

Then, we have the final equations

$$\frac{\rho^2 A r_0^2}{EI\kappa G}\omega^4 - 2\frac{\rho^2 A r_0^2 \Omega}{EI\kappa G}\omega^3 - \frac{1}{EI}\left[\left(\frac{n\pi}{L}\right)^2\left(\frac{EI\rho}{\kappa G} + \rho A r_0^2\right) + \rho A\right]\omega^2$$

$$+\frac{1}{EI}\left(\frac{n\pi}{L}\right)^2 2\rho A r_0^2 \Omega\omega + \left(\frac{n\pi}{L}\right)^4 = 0 \tag{A.27}$$

For $\Omega = 0$, the frequency equation can be simplified

$$\frac{\rho^2 A r_0^2}{EI\kappa G}\omega^4 - \frac{1}{EI}\left[\left(\frac{n\pi}{L}\right)^2\left(\frac{EI\rho}{\kappa G} + \rho A r_0^2\right) + \rho A\right]\omega^2 + \left(\frac{n\pi}{L}\right)^4 = 0 \tag{A.28}$$

Finally the eigenfrequencies of the undamaged shaft can be determined

$$
\omega_i^2 = \frac{\frac{1}{EI}\left[\left(\frac{n\pi}{L}\right)^2\left(\frac{EI\rho}{\kappa G} + \rho A r_0^2\right) + \rho A\right]}{2\frac{\rho^2 A r_0^2}{EI\kappa G}}
$$

$$
\pm \frac{\sqrt{\frac{1}{(EI)^2}\left[\left(\frac{n\pi}{L}\right)^2\left(\frac{EI\rho}{\kappa G} + \rho A r_0^2\right) + \rho A\right]^2 - 4\frac{\rho^2 A r_0^2}{EI\kappa G}\left(\frac{n\pi}{L}\right)^4}}{2\frac{\rho^2 A r_0^2}{EI\kappa G}} \tag{A.29}
$$

A.2 Dunkerley's equation

The Dunkerley equation gives the lower bound of the fundamental frequency of vibration of a composite system in terms of the frequencies of vibration of the system's partial systems. The influence coefficient is linear (angular) deflection of the point due to the unit force (moment) being applied at the same point. Consider the free vibration of an undamped system. By the method of influence coefficients, we have

$$\{q\} = [d_{ij}]\{-m\ddot{q}\} \tag{A.30}$$

where $\{q\}$ is the displacement vector, $[d_{ij}]$ the flexibility matrix, and $\{-m\ddot{q}\}$ the vector of inertia forces. At a principal mode of vibration, the deflections $\{q\}$ are harmonics with $\{\ddot{q}\} = \{-\omega^2 q\}$. Substituting this in the Eq.(A.30) gives

$$\{q\} = [d_{ij}]\{\omega^2 mq\} \tag{A.31}$$

Let us illustrate the method for a two-degree-of-freedom system, from Eq.(A.31) we have

$$\left\{ \begin{array}{c} q_1 \\ q_2 \end{array} \right\} = \left[\begin{array}{cc} d_{11} & d_{12} \\ d_{21} & d_{22} \end{array} \right] \left\{ \begin{array}{c} \omega^2 m_1 q_1 \\ \omega^2 m_2 q_2 \end{array} \right\} \tag{A.32}$$

Dividing by ω^2 and rearranging, we obtain

$$\left[\begin{array}{cc} \frac{1}{\omega^2} - d_{11}m_1 & -d_{12}m_2 \\ -d_{21}m_1 & \frac{1}{\omega^2} - d_{22}m_2 \end{array} \right] \left\{ \begin{array}{c} q_1 \\ q_2 \end{array} \right\} = \left\{ \begin{array}{c} 0 \\ 0 \end{array} \right\} \tag{A.33}$$

Hence the frequency equation is

$$\left| \begin{array}{cc} \frac{1}{\omega^2} - d_{11}m_1 & -d_{12}m_2 \\ -d_{21}m_1 & \frac{1}{\omega^2} - d_{22}m_2 \end{array} \right| = 0 \tag{A.34}$$

or

$$\left(\frac{1}{\omega^2}\right)^2 - (d_{11}m_1 + d_{22}m_2)\left(\frac{1}{\omega^2}\right) + m_1 m_2 (d_{11}d_{22} - d_{12}d_{21}) = 0 \tag{A.35}$$

Assume ω_1 and ω_2 are the natural frequencies. The factored form of Eq.(A.35) yields

$$\left(\frac{1}{\omega^2} - \frac{1}{\omega_1^2}\right)\left(\frac{1}{\omega^2} - \frac{1}{\omega_2^2}\right) = 0 \tag{A.36}$$

or

$$\left(\frac{1}{\omega^2}\right)^2 - \left(\frac{1}{\omega_1^2} + \frac{1}{\omega_2^2}\right)\left(\frac{1}{\omega^2}\right) + \frac{1}{\omega_1^2\omega_2^2} = 0 \tag{A.37}$$

Equating the coefficients of the $\frac{1}{\omega^2}$ terms in Eqs.(A.35) and (A.37) yields

$$\frac{1}{\omega_1^2} + \frac{1}{\omega_2^2} = d_{11}m_1 + d_{22}m_2 \tag{A.38}$$

If the fundamental frequency ω_1 is much lower than that of the harmonic ω_2, we have $\frac{1}{\omega_2^2} \ll \frac{1}{\omega_1^2}$ and

$$\frac{1}{\omega_1^2} \cong d_{11}m_1 + d_{22}m_2 \tag{A.39}$$

Eq.(A.39) is called as Dunkerley's equation. The corresponding equation for a multi-degree-of-freedom system is

$$\frac{1}{\omega_1^2} \cong d_{11}m_1 + d_{22}m_2 + \cdots + d_{nn}m_n = \sum_{i=1}^{n} d_{ii}m_i \tag{A.40}$$

The influence coefficient d_{ii} is the deflection at station i of the system due to a unit force applied at the same location. The quantity $d_{ii}m_i$ is due to m_i acting alone, that is, the inertial effect of all other masses is not considered. Since d_{ii} is the reciprocal of the stiffness k_{ii} we have

$$d_{ii}m_i = \frac{m_i}{k_{ii}} = \frac{1}{\omega_{ii}^2} \tag{A.41}$$

where ω_{ii} is the natural frequency of an equivalent mass-spring system with m_i acting alone at station i. Hence an alternative form of Dunkerley's equation is the relationship between the fundamental frequency of the actual system and partial frequencies

$$\frac{1}{\omega_1^2} \geqslant \frac{1}{\omega_{11}^2} + \frac{1}{\omega_{22}^2} + \cdots + \frac{1}{\omega_{nn}^2} = \sum_{i=1}^{n} \frac{1}{\omega_{ii}^2} \tag{A.42}$$

Note that the estimated fundamental frequency is always lower than the exact value, since the harmonics are neglected in the equation.

Critical speed due to deflection from shaft weight only

$$\omega_s^2 = \pi^4 \frac{EI}{m_s L^3} \approx 97.417 \frac{EI}{\rho A L^4} \tag{A.43}$$

Critical speed due to deflection from load (disk) only

$$k = \frac{48EI}{L^3} \tag{A.44}$$

$$\omega_d^2 = \sqrt{\frac{k}{m_d}} = \sqrt{48 \frac{EI}{m_d L^3}} \tag{A.45}$$

Critical speed for rotor

$$\omega_1^2 \geq \frac{1}{D} \tag{A.46}$$

$$D = \sum_i \frac{1}{\omega_i^2} = \frac{1}{\omega_s^2} + \frac{1}{\omega_d^2} = \frac{1}{\frac{97.417EI}{\rho A L^4}} + \frac{1}{\frac{48EI}{m_d L^3}} \tag{A.47}$$

$$\omega_1^2 \geq \frac{EI}{L^3 \left(\frac{\rho A L}{97.417} + \frac{m_d}{48} \right)} \tag{A.48}$$

A.3 Rayleigh's method

Rayleigh's method assumes that the system is conservative, and at a principal mode, the maximum potential energy of the system is equal to its maximum kinetic energy. It is necessary to assume the dynamic mode shape, or the modal vector, in order to estimate the natural frequencies. Generally, Rayleigh's method is used to find the fundamental frequency, since the modal vectors for the higher frequencies are more difficult to estimate. If an exact mode shape is assumed, the frequency calculated will be exact. If the assumed mode shape is not the exact dynamic mode shape, it is equivalent to the application of additional constraints to the vibratory system. Hence the calculated frequency is higher than the true value. Thus, Rayleigh's method tends to give a higher value for the estimated frequency.

Usually the dynamic deflections are estimated from the static deflections. The potential energy of the system is the strain energy in the bent shaft, which is the work done by static loads. The maximum potential energy U_{max} is

$$U_{max} = \frac{1}{2}\left(m_1 y_1 + m_2 y_2 + \cdots + m_n y_n\right) g = \frac{1}{2} g \sum_{i=1}^{n} m_i y_i \tag{A.49}$$

where $m_i g$ is the static load due to a rotor and y_i is the deflection at the rotor. For harmonic oscillation, the maximum kinetic energy T_{max} due to rotors is

$$T_{max} = \frac{1}{2}\omega^2 \left(m_1 y_1^2 + m_2 y_2^2 + \cdots + m_n y_n^2\right) g = \frac{1}{2}\omega^2 \sum_{i=1}^{n} m_i y_i^2 \tag{A.50}$$

ω is the frequency of oscillation. Equating U_{max} and T_{max} and simplifying, we obtain

$$\omega^2 = \frac{g\left(m_1 y_1 + m_2 y_2 + \cdots + m_n y_n\right)}{\left(m_1 y_1^2 + m_2 y_2^2 + \cdots + m_n y_n^2\right)} = g\frac{\sum_{i=1}^{n} m_i y_i}{\sum_{i=1}^{n} m_i y_i^2} \tag{A.51}$$

Note that the equation is derived from the lateral beam deflection of a shaft. The frequency for the transverse vibration of the system is also the critical speed at which the shaft whirl takes place.

For a continuous shaft system where the mass of the shaft is not ignored, the maximum potential of the shaft can be written as

$$U_{max} = \int_0^L EI \left(\frac{\mathrm{d}^2 y}{\mathrm{d}z^2}\right)^2 \mathrm{d}z \tag{A.52}$$

the maximum kinetic energy T_{max} of the shaft is

$$T_{max} = \int_0^L (\omega y)^2 \, \mathrm{d}m = \omega^2 \int_0^L \rho A \left(y\left(z\right)\right)^2 \mathrm{d}z \tag{A.53}$$

Hence the Rayleigh quotient for shaft vibration is

$$\omega^2 = \frac{\int_0^L EI \left(\frac{\mathrm{d}^2 y}{\mathrm{d}z^2}\right)^2 \mathrm{d}z}{\int_0^L \rho A \left(y\left(z\right)\right)^2 \mathrm{d}z + \sum m_i \left(y_i\left(z\right)\right)^2} \tag{A.54}$$

A mode shape function is assumed to be

$$y\left(z\right) = y_0 \sin\left(\frac{\pi z}{L}\right) \tag{A.55}$$

which satisfies the boundary conditions: $y\left(0\right) = y\left(L\right) = 0$ and $y''\left(0\right) = y''\left(L\right) = 0$.

Potential and kinetic energy of shaft

$$U_s = \frac{1}{2}\int_0^L EI \left(\frac{\mathrm{d}^2 y}{\mathrm{d}z^2}\right)^2 \mathrm{d}z = \frac{1}{4} y_0^2 \frac{\pi^4 EI}{L^3} \tag{A.56}$$

$$T_s = \frac{1}{2}\int_0^L z^2 \, \mathrm{d}m = \frac{1}{4} y_0^2 m_s \tag{A.57}$$

Kinetic energy of disk

$$T_d = \frac{1}{2} m_d y_{\left(z=\frac{L}{2}\right)}^2 = \frac{1}{2} m_d y_0^2 \tag{A.58}$$

Rayleigh-Quotient:

$$\begin{aligned}
\omega_1^2 \;&\leq\; \frac{U_s}{T_s + T_d} = \frac{\frac{1}{4} y_0^2 \frac{\pi^4 EI}{L^3}}{\frac{1}{4} y_0^2 m_s + \frac{1}{2} y_0^2 m_d} \\
&\leq\; \frac{\pi^4}{m_s + 2m_d} \frac{EI}{L^3}
\end{aligned} \tag{A.59}$$

List of Figures

1.1 Physical phenomena leading to rotor cracks [91] 1
1.2 Research design 13
1.3 Structure of the dissertation and corresponding chapters 16

2.1 Fracture process zone model 21
2.2 Traction separation law for brittle materials 22
2.3 Fracture process in cohesive zone model 23
2.4 Representation of the fracture process using CZM in FE model 24
2.5 Simulation of one cohesive element 25
2.6 Bilinear traction separation law written in subroutine 26
2.7 Finite element results of one cohesive element for different values of parameters, σ_{max}, G_{IC}, K_p and δ 27
2.8 Thin steel plate contains an edge crack subjected to normal stress 28
2.9 Elastic stress distributions at the crack tip 29
2.10 Stress distribution of Irwin's model ahead of the crack tip 30
2.11 Dugdale's approach ahead of the crack tip 30
2.12 Schematic of an edge crack using CZM in FE model 31
2.13 Stress distributions in front of the crack tip using LEFM, EPFM and CZM [82] 32
2.14 Normalised stress intensity factor versus normalised applied stress 32
2.15 Numerical simulation and experimental data [142] for a Mode-I DCB test and mixed mode I-II MMB test 35
2.16 Geometry of the crack growth using cohesive elements 38
2.17 Variation of elastic strain energy and cohesive energy for pure elastic material 40
2.18 Variation of elastic strain energy, plastic dissipative energy and cohesive energy for elasto-plastic material 40
2.19 Variation of plastic energy and cohesive energy for different σ_{max}/σ_Y ratio 41
2.20 Variation of plastic energy and cohesive energy for different strain hardening exponent n 41
2.21 Triaxiality parameter χ at initiation of ductile fracture in different geometries 42
2.22 Visualization of the two approaches used for description of the bonding region 43
2.23 Traction separation law for ductile materials 44
2.24 Dependence of triaxiality parameter on a fixed applied stress ratio under plain strain 47
2.25 TSL shape proposed by Scheider [118] 48
2.26 TSL shape proposed by Scheider [118] compared to Banerjee [13] 49
2.27 Traction-separation for different stress ratios r_σ for steel and aluminium . . 51

2.28 Effect of stress ratios r_σ on peak stress for steel 52

2.29 Dependence of the normalised cohesive parameters: cohesive strength σ_{max}/σ_Y and cohesive energy G_I/σ_Y on the triaxiality for steel 52

2.30 Breathing crack model of the non rotating cracked shaft subjected to rotating load 54

2.31 Breathing crack of the non rotating cracked shaft under rotating load for relative crack depth a/d=0.1 55

2.32 Variation of function $F(a/d)$ vs. relative crack depth a/d 56

2.33 Rotor with an open crack simply supported on both ends 56

2.34 Dimensionless flexibility of the cracked section for load direction normal to crack edge 60

3.1 Stationary and rotating coordinate system of the cracked shaft 62

3.2 Open crack model and various breathing function models 65

3.3 Comparison between open, Mayes' and Yang's crack model for different relative crack depth a/d 66

3.4 Crack state variations during rotation: perpendicular to crack front line model [30], [66], [136] 67

3.5 Crack state variations during rotation: elliptical crack model [131], [11] . . 67

3.6 Crack state variations during rotation suggested by Lees and Friswell [77] . 68

3.7 Crack state variations during rotation represented by Al-Shudeifat and Butcher [2] 69

3.8 Crack state variations during rotation: parallel to crack front line model [80], [83] 70

3.9 Crack state variations during rotation reported by Bachschmid et al. [10] . 71

3.10 Crack in early propagation state [11] 71

3.11 Deep propagated crack [11] 72

3.12 Experimentally determined crack propagation patterns [86], [11] 72

3.13 Crack closure parabolic line model for various relative crack depths a/d . . 73

3.14 Model of shaft with a transverse crack 75

3.15 Geometry of the cracked shaft 76

3.16 Geometry of the crack closure straight line model 79

3.17 Breathing condition under rotation due to bending moment of gravity . . . 80

3.18 Normalised shaft stiffness variations of a cracked rotor with a breathing crack for relative crack depth $a/d = 0.1$ 83

3.19 Curve fitting of normalised shaft stiffness for relative crack depth $a/d = 0.1$ 84

3.20 Normalised shaft stiffness vs. various relative crack depths a/d in rotating coordinates 85

3.21 Normalised shaft stiffness vs. various relative crack depths a/d in fixed and rotating coordinates 86

3.22 Vibration response in rotating coordinates for $a/d = 0.1$, $\Omega = 500$ rad/s . . 87

3.23 Vibration response in rotating coordinates for $a/d = 0.1$, $\Omega = 1000$ rad/s . 87

3.24 Steady state orbital responses in rotating coordinates for relative crack depth $a/d = 0.1$, $\Omega = 500$ rad/s and 1000 rad/s 88

3.25 Geometry of crack closure parabolic line of the breathing cracked shaft . . 89

3.26 Normalised shaft stiffnesses for relative crack depth $a/d = 0.1 \div 0.4$ 92

4.1 Approximations to the boundary between stability and instability of the linear-coupled Mathieu equations for $n=0, 1$ and 2 102

4.2 Borderlines of stability with a damping of 5% 102

5.1 Rotor with a single rigid disk supported on both ends and coordinate system 104

5.2 Geometry of cylindrical shaft element 105

5.3 Geometry of rigid disk 110

5.4 Cross section of the cracked shaft 111

5.5 Assembling of the complete cracked rotor system 113

5.6 Assembling of the cracked rotor system: first proposed FE model 114

5.7 Natural frequencies for the uncracked and cracked rotor without disk . . . 116

5.8 Variation of mode shapes of the uncracked and cracked rotor for the first three natural frequencies for relative crack depth a/d=0.1 and 0.4 117

5.9 First natural frequency ratio for different crack depths 118

5.10 First of three natural frequencies for different crack depths of shaft without disk for relative crack depth a/d from 0.1 to 0.4 119

5.11 Traction-separation law in FE computation 122

5.12 Assembling of the cracked rotor system: second proposed FE model 122

5.13 Natural frequencies for the cracked rotor without disk based on Mayes' model, and two proposed FE models 124

5.14 Natural frequencies for the uncracked and cracked rotor with disk 126

6.1 Breathing crack with weight dominance [21] 129

6.2 Principle of the simulation using an integration of FE and MBS [81] 130

6.3 Finite element model of a flexible cracked shaft 131

6.4 Normalised natural frequencies versus relative crack depth 132

6.5 Comparison of the mode shapes between cracked and uncracked shaft . . . 132

6.6 FE model of the elastic cracked shaft supported by rigid bearings in MBS . 133

6.7 Deflection line due to contributions of the uncracked shaft and the local crack compliance 134

6.8 FE model of elastic cracked shafts loaded by weight only 135

6.9 Vibration amplitude of flexible cracked shaft loaded by weight only at various rotating speeds 135

6.10 Steady state orbital responses of flexible cracked shaft loaded by weight only at various rotating speeds 136

6.11 FE model of elastic uncracked and cracked shaft with unbalance in MBS . 137

6.12 Geometry of cracked shaft model for relative crack depth a/d=0.5 with large unbalance mass 137

6.13 Vibration amplitude of flexible uncracked and cracked shaft loaded by weight and large unbalance at various rotating speeds: relative crack depth a/d=0.5 138

6.14 Vibration amplitude of flexible uncracked and cracked shaft loaded by weight and small unbalance at various rotating speeds: relative crack depth a/d=0.5 140

6.15 Vibration amplitude of flexible uncracked and shallow cracked shaft loaded by weight and large unbalance at various rotating speeds: relative crack depth a/d=0.1 140

6.16 Evaluation of displacement and stress criteria for state of crack 141

6.17 Breathing crack of the non rotating cracked shaft under rotating load for relative crack depth a/d=0.10 143

6.18 Breathing crack of the non rotating cracked shaft under rotating load for relative crack depth a/d=0.20 144

6.19 Export the vibration responses from MBS into FE 145

6.20 MBS result: Lateral vibration responses of a cracked rotor during rotation 146

6.21 MBS result: Steady state orbital responses of a cracked rotor during rotation 146

6.22 Breathing crack of the rotating cracked shaft for relative crack depth a/d=0.10 147

6.23 Rotating cracked shaft during one revolution 149

6.24 Comparison between the crack closure straight line model and the simulation results of breathing mechanism for relative crack depth a/d=0.1 152

List of Tables

2.1 Mechanical properties of steel 36
2.2 Mechanical properties of cohesive element 36
2.3 Energy balance for a single-edge crack plate 37
2.4 Natural frequencies of uncracked and cracked shaft a/d=0.1 at rest in [rad/s] 57
2.5 Natural frequencies of uncracked shaft at rest in [rad/s] 60

3.1 Crack states for half angle rotation $\Omega t = 0 \div \pi$ for $a/d = 0.2$ 90

5.1 Shaft parameters 115
5.2 Natural frequencies of uncracked shaft without disk at rest in [Hz] 120
5.3 Natural frequencies of cracked shaft without disk at rest in [Hz] 120
5.4 First natural frequencies of cracked shaft without disk in [Hz] 123
5.5 Second natural frequencies of cracked shaft without disk in [Hz] 123
5.6 Disk parameters 125
5.7 Natural frequencies of uncracked shaft with disk in [Hz] 127
5.8 Natural frequencies of cracked shaft with disk in [Hz] 127

6.1 Comparison simulation results of breathing mechanism between rotating shaft and rotating load for relative crack depth a/d=0.1 at rotating speed Ω=600 rpm 148

Bibliography

[1] M. L. Adams. *Rotating machinery vibration: From analysis to troubleshooting.* CRC Press, Taylor and Francis Group, 2010.

[2] M. A. Al-Shudeifat und E. A. Butcher. *New breathing functions for the transverse breathing crack of the cracked rotor system: Approach for critical and subcritical harmonic analysis. Journal of Sound and Vibration*, 330:526–544, 2011.

[3] M. A. Al-Shudeifat, E. A. Butcher und C. R. Stern. *General harmonic balance solution of a cracked rotor-bearing-disk system for harmonic and sub-harmonic analysis: Analytical and experimental approach. International Journal of Engineering Science*, 48:921–935, 2010.

[4] G. Alvano und E. Sacco. *Combining interface damage and friction in a cohesive-zone model. International Journal for Numerical Methods in Engineering*, 68:542–582, 2006.

[5] S. Andrieux und C. Varé. *A 3D cracked beam model with unilateral contact. Application to rotors. European Journal of Mechanics-A/Solids*, 21:793–810, 2002.

[6] M. Anvari, I. Scheider und C. Thaulow. *Simulation of dynamic ductile crack growth using strain-rate and triaxiality-dependent cohesive elements. Engineering Fracture Mechanics*, 73:2210–2228, 2006.

[7] S. E. Arem. *Shearing effects on the breathing mechanism of a cracked beam section in bi-axial flexure. European Journal of Mechanics-A/Solids*, 28(6):1079–1087, 2009.

[8] S. E. Arem und H. Maitournam. *A cracked beam finite element for rotating shaft dynamics and stability analysis. Journal of Mechanics of Materials and Structures*, 3(5):893–910, 2008.

[9] N. Bachschmid und E. Tanzi. *Deflections and strains in cracked shafts due to rotating loads: a numerical and experimental analysis. International Journal of Rotating Machinery*, 10(4):283–291, 2004.

[10] N. Bachschmid, P. Pennacchi und E. Tanzi. *Some remarks on breathing mechanism, on non-linear effects and on slant and helicoidal cracks. Mechanical Systems and Signal Processing*, 22:879–904, 2008.

[11] N. Bachschmid, P. Pennacchi und E. Tanzi. *On the evolution of vibrations in cracked rotors. In Proceeding of the 8th IFToMM International Conference on Rotor Dynamics, KIST, Seoul, Korea*, 2010.

[12] N. Bachschmid, P. Pennacchi und E. Tanzi. *Cracked rotors*. Springer-Verlag, 2010.

[13] A. Banerjee und R. Manivasagam. *Triaxiality dependent cohesive zone model*. *Engineering Fracture Mechanics*, 76:1761–1770, 2009.

[14] G. I. Barenblatt. *The mathematical theory of equilibrium crack in brittle fracture*. *Advances of Applied Mechanics*, 7:55–129, 1962.

[15] A. Bouboulas und N. Anifantis. *Finite element modeling of a vibrating beam with a breathing crack: observations on crack detection*. In *Structural Health Monitoring*, 2010.

[16] W. Brocks. *Cracks and Plasticity*. Technical report, Institut für Werkstoffforschung, GKSS-Forschungszentrum Geesthacht, 2008.

[17] D. Broek. *Elementary engineering fracture mechanics*. Martinus Nijhoff Publishers, 1984.

[18] A. C. Chasalevris und C. A. Papadopoulos. *Coupled horizontal and vertical bending vibrations of a stationary shaft with two cracks*. *Journal of Sound and Vibration*, 309:507–528, 2008.

[19] A. C. Chasalevris und C. A. Papadopoulos. *A continuous model approach for cross-coupled bending vibrations of a rotor-bearing system with a transverse breathing crack*. *Mechanism and Machine Theory*, 44:1176–1191, 2009.

[20] L.-W. Chen und H.-K. Chen. *Stability analyses of a cracked shaft subjected to the end load*. *Journal of Sound and Vibration*, 188(4):497–513, 1995.

[21] L. Cheng, N. Li, X.-F. Chen und Z.-J. He. *The influence of crack breathing and imbalance orientation angle on the characteristics of the critical speed of a cracked rotor*. *Journal of Sound and Vibration*, 330:2031–2048, 2011.

[22] D. Childs. *Turbomachinery rotordynamics: Phenomena, modeling, and analysis*. John Wiley and Sons, Inc., 1993.

[23] T. G. Chondros, A. D. Dimarogonas und J. Yao. *A continuous cracked beam vibration theory*. *Journal of Sound and Vibration*, 215(1):17–34, 1998.

[24] T. G. Chondros, A. D. Dimarogonas und J. Yao. *Vibration of a beam with a breathing crack*. *Journal of Sound and Vibration*, 239(1):57–67, 2001.

[25] A. Cornec, I. Scheider und K.-H. Schwalbe. *On the practical application of the cohesive model*. *Engineering Fracture Mechanics*, 70:1963–1987, 2003.

[26] B. Cotterell. *The past, present, and future of fracture mechanics*. *Engineering Fracture Mechanics*, 69:533–553, 2002.

[27] G. Curti, F. A. Raffa und F. Vatta. *An analytical approach to the dynamics of rotating shafts*. *Meccanica*, 27:285–292, 1992.

[28] M. H. F. Dado und O. Abuzeid. *Coupled transverse and axial vibratory behaviour of cracked beam with end mass and rotary inertia. Journal of Sound and Vibration,* 261:675–696, 2003.

[29] O. E. K. Daoud und D. J. Cartwright. *Strain energy release rates for a straight-fronted edge crack in a circular bar subject to bending. Engineering Fracture Mechanics,* 19(4):701–707, 1984.

[30] A. K. Darpe, K. Gupta und A. Chawla. *Coupled bending, longitudinal and torsional vibrations of a cracked rotor. Journal of Sound and Vibration,* 269:33–60, 2004.

[31] A. K. Darpe, K. Gupta und A. Chawla. *Transient response and breathing behaviour of a cracked Jeffcott rotor. Journal of Sound and Vibration,* 272:207–243, 2004.

[32] N. Dharmaraju, R. Tiwari und S. Talukdar. *Identification of an open crack model in a beam based on force-response measurements. Computers and Structures,* 82: 167–179, 2004.

[33] A. D. Dimarogonas. *Vibration engineering.* St. Paul: West Publishers, 1976.

[34] A. D. Dimarogonas. *Vibration of cracked structures: A state of the art review. Engineering Fracture Mechanics,* 55(5):831–857, 1996.

[35] A. D. Dimarogonas und C. A. Papadopoulos. *Vibration of cracked shafts in bending. Journal of Sound and Vibration,* 91(4):583–593, 1983.

[36] G. M. Dong, J. Chen und J. Zou. *Parameter identification of a rotor with an open crack. European Journal of Mechanics-A/Solids,* 23:325–333, 2004.

[37] D. S. Dugdale. *Yielding of steel sheets containing slits. Journal of Mechanics and Physics Solid,* 8:100–104, 1960.

[38] M. Elices, G. V. Guinea, J. Gómez und J. Planas. *The cohesive zone model: advantages, limitations and challenges. Engineering Fracture Mechanics,* 69:137–163, 2002.

[39] Z. Friedman und J. Kosmatka. *An improved two-node Timoshenko beam finite element. Computers and Structures,* 47(3):473–481, 1993.

[40] R. Gasch. *Dynamic behaviour of a simple rotor with a cross-sectional crack.* In *Proceeding of the IME Conference Publication Vibrations in Rotating Machinery, Paper No. C178/76.,* 1976.

[41] R. Gasch. *A survey of the dynamic behaviour of a simple rotating shaft with a transverse crack. Journal of Sound and Vibration,* 160(2):313–332, 1993.

[42] R. Gasch. *Dynamic behavior of the Laval rotor with a transverse crack. Mechanical Systems and Signal Processing,* 22:790–804, 2008.

[43] R. Gasch und H. Pfützner. *Rotordynamik: Eine Einführung.* Springer-Verlag, 1975.

[44] R. Gasch, R. Nordmann und H. Pfützner. *Rotordynamik, 2. Auflage.* Springer-Verlag, 2001.

[45] E. E. Gdoutos. *Fracture mechanics: An introduction, 2nd Edition.* Springer-Verlag, 2005.

[46] G. Genta. *Consistent matrices in rotor dynamics. Meccanica,* 20:235–248, 1985.

[47] S. K. Georgantzinos und N. K. Anifantis. *An insight into the breathing mechanism of a crack in a rotating shaft. Journal of Sound and Vibration,* 318:279–295, 2008.

[48] P. H. Geubelle und J. S. Baylor. *Impact-induced delamination of composites: a 2D simulation. Composites Part B,* 29B:589–602, 1998.

[49] B. Grabowski. *The vibrational behavior of a turbine rotor containing a transverse crack.* In *Proceeding of the ASME Design Engineering Technology Conference, St. Louis, Paper No.79-DET-67.,* 1979.

[50] D. Gross und T. Seelig. *Bruchmechanik: Mit einer Einführung in die Mikromechanik, 4. Auflage.* Springer-Verlag, 2007.

[51] A. Gurson. *Continuum theory of ductile rupture by void nucleation and growth: part I - yield criteria and flow rules for porous ductile media. Journal of Engineering Materials and Technology - Trans ASME,* 99:2–15, 1977.

[52] P. Hagedorn. *Nichtlineare Schwingungen.* Akademische Verlagsgesellschaft Wiesbaden, 1978.

[53] J. W. Hancock und D. K. Brown. *On the role of strain and stress state in ductile failure. Journal of the Mechanics and Physics of Solids,* 31(1):1–24, 1983.

[54] J. W. Hancock und A. C. Mackenzie. *On the mechanisms of ductile failure in high-strength steels subjected to multi-axial stress-states. Journal of the Mechanics and Physics of Solids,* 24:147–169, 1976.

[55] J. Hansen. *Stability diagrams for coupled Mathieu-equations. Ingenieur-Archiv,* 55: 463–473, 1985.

[56] A. Hillerborg, M. Modéer und P. Petersson. *Analysis of crack formation and crack growth in concrete by means of fracture mechanics and finite elements. Cement and Concrete Research,* 6:773–782, 1976.

[57] G. Härkegard, J. Denk und K. Stärk. *Growth of naturally initiated fatigue cracks in ferritic gas turbine rotor steels. International Journal of Fatigue,* 27:715–726, 2005.

[58] S. C. Huang, Y. M. Huang und S. M. Shieh. *Vibration and stability of a rotating shaft containing a transverse crack. Journal of Sound and Vibration,* 162(3):387–401, 1993.

[59] J. W. Hutchinson und A. G. Evans. *Mechanics of materials: Top-down approaches to fracture.* Acta Materialia, 48:125–135, 2000.

[60] T. Ikeda und S. Murakami. *Dynamic response and stability of a rotating asymmetric shaft mounted on a flexible base.* Nonlinear Dynamics, 20:1–19, 1999.

[61] G. Irwin und J. Kies. *Fracturing and fracture dynamics.* Welding Journal Research Supplement, 31(2):95s–100s, 1952.

[62] Y. Ishida. *Cracked rotors: Industrial machine case histories and nonlinear effects shown by simple Jeffcott rotor.* Mechanical Systems and Signal Processing, 22:805–817, 2008.

[63] Z.-H. Jin und C. T. Sun. *A comparison of cohesive zone modeling and classical fracture mechanics based on near tip stress field.* International Journal of Solids and Structures, 43:1047–1060, 2006.

[64] O. S. Jun und M. S. Gadala. *Dynamic behavior analysis of cracked rotor.* Journal of Sound and Vibration, 309:210–245, 2008.

[65] O. S. Jun und J. O. Kim. *Free bending vibration of a multi-step rotor.* Journal of Sound and Vibration, 224(4):625–642, 1999.

[66] O. S. Jun, H. J. Eun, Y. Y. Earmme und C.-W. Lee. *Modelling and vibration analysis of a simple rotor with a breathing crack.* Journal of Sound and Vibration, 155(2): 273–290, 1992.

[67] T. Kaneko. *On Timoshenko's correction for shear in vibrating beams.* Journal of Physics D: Applied Physics, 8:1927–1936, 1975.

[68] T. Kaneko. *An experimental study of Timoshenko's shear coefficient for flexurally vibrating beams.* Journal of Physics D: Applied Physics, 11:1979–1988, 1978.

[69] I. A. Karnovsky und O. I. Lebed. *Formulas for structural dynamics.* McGraw-Hill Co., 2004.

[70] H. Keiner und M. S. Gadala. *Comparison of different modelling techniques to simulate the vibration of a cracked rotor.* Journal of Sound and Vibration, 254(5): 1012–1024, 2002.

[71] M. L. Kikidis und C. A. Papadopoulos. *Slenderness ratio effect on cracked beam.* Journal of Sound and Vibration, 155(1):1–11, 1992.

[72] M. Kisa und J. Brandon. *The effects of closure of cracks on the dynamics of a cracked cantilever beam.* Journal of Sound and Vibration, 238(1):1–18, 2000.

[73] M. Kisa, J. Brandon und M. Topcu. *Free vibration analysis of cracked beams by a combination of finite elements and component mode synthesis methods.* Computers and Structures, 67:215–223, 1998.

[74] C. Kumar und V. Rastogi. *A brief review on dynamics of a cracked rotor. International Journal of Rotating Machinery*, 2009.

[75] M. Lalanne und G. Ferraris. *Rotordynamics prediction in engineering*. John Wiley and Sons, 1990.

[76] C.-W. Lee. *Vibration analysis of rotors*. Kluwer Academic, 1993.

[77] A. W. Lees und M. Friswell. *Crack detection in asymmetric rotors. Key Engineering Materials*, 167-168:246–255, 1999.

[78] J. Lemaitre und R. Desmorat. *Engineering damage mechanics: Ductile, creep, fatigue and brittle failures*. Springer-Verlag, 2005.

[79] H. Li und N. Chandra. *Analysis of crack growth and crack-tip plasticity in ductile materials using cohesive zone models. International Journal of Plasticity*, 19:849–882, 2003.

[80] R. T. Liong und C. Proppe. *Application of a cohesive zone model for the investigation of the dynamic behaviour of a rotating shaft with a transverse crack.* In *Proceeding of the 8th IFToMM International Conference on Rotor Dynamics, KIST, Seoul, Korea*, 2010, 628-636.

[81] R. T. Liong und C. Proppe. *Application of the cohesive zone model to the analysis of a rotor with a transverse crack.* In *Proceeding of the 8th International Conference on Structural Dynamic EURODYN2011, Leuven, Belgium*, 2011, 3434-3442.

[82] R. T. Liong und C. Proppe. *Implementation of a cohesive zone model for crack closure analysis of rotors. PAMM Wiley Inter Science-GAMM2009 Gdansk-Poland*, 9:181–182, 2009.

[83] R. T. Liong und C. Proppe. *Implementation of a cohesive zone model for the investigation of the dynamic behaviour of a rotating shaft with a transverse crack. PAMM Wiley Inter Science-GAMM2010 Karlsruhe, Germany*, 10:125–126, 2010.

[84] R. T. Liong und C. Proppe. *Cohesive zone model for a transverse breathing crack in a rotor. PAMM Wiley Inter Science-GAMM2011 Graz-Austria*, 11:163–164, 2011.

[85] G. Litak und J. T. Sawicki. *Intermittent behaviour of a cracked rotor in the resonance region. Chaos, Solitons and Fractals*, 42:1495–1501, 2009.

[86] T. Lorentzen, N. E. Kjaer und T. K. Henriksen. *The application of fracture mechanics to surface cracks in shafts. Engineering Fracture Mechanics*, 23(6):1005–1014, 1986.

[87] G. M. Mahmoud. *Stability regions for coupled Hill's equations. Journal of Physics A*, 242:239–249, 1997.

[88] J. G. Mancilla und J. M. M. Lopez. *Crack breathing mechanisms in rotor-bearings systems, its influence on system response and crack detection.* In *Proceeding of the ISCORMA-3, Cleveland, Ohio*, 2005.

[89] I. Mayes und W. Davies. *The vibrational behaviour of a rotating system containing a transverse crack*. In *Proceeding of the MechE Conference on Vibrations in Rotating Machinery, C/168/6*, 1976.

[90] I. Mayes und W. Davies. *Analysis of the response of a multi-rotor bearing system containing a transverse crack in a rotor*. Journal of Vibration, Acoustics, Stress, and Reliability in Design, 106:139–145, 1984.

[91] A. Muszynska. *Rotordynamics*. Taylor and Francis Group, 2005.

[92] A. Muszynska, P. Goldman und D. Bently. *Torsional/lateral cross-coupled responses due to shaft anisotropy: a new tool in shaft crack detection*. In *Proceeding of the Vibration in Rotating Machinery, Institution of Mechanical Engineers Conference Publications, London, England*, 1992.

[93] A. H. Nayfeh und D. T. Mook. *Nonlinear oscillations*. John Wiley and Sons, Inc., 1979.

[94] A. Needleman. *An analysis of tensile decohesion along an interface*. Journal of the Mechanics and Physics of Solids, 38(3):289–324, 1990.

[95] A. Needleman und V. Tvergaard. *An analysis of ductile rupture modes at a crack tip*. Journal of the Mechanics and Physics of Solids, 35(2):151–183, 1987.

[96] F. C. Nelson. *Rotor dynamics without equations*. International Journal of COMA-DEM, 10(3):2–10, 2007.

[97] O. Nguyen, E. A. Repetto, M. Ortiz und R. A. Radovitzky. *A cohesive model of fatigue crack growth*. International Journal of Fracture, 110:351–369, 2001.

[98] W. M. Ostachowicz und M. Krawczuk. *Coupled torsional and bending vibrations of a rotor with an open crack*. Archive of Applied Mechanics, 62:191–201, 1992.

[99] C. A. Papadopoulos. *The strain energy release approach for modeling cracks in rotors: A state of the art review*. Mechanical Systems and Signal Processing, 22: 763–789, 2008.

[100] C. A. Papadopoulos und A. D. Dimarogonas. *Coupling of bending and torsional vibration of a cracked Timoshenko shaft*. Archive of Applied Mechanics (Ingenieur-Archiv), 57:257–266, 1987.

[101] C. A. Papadopoulos und A. D. Dimarogonas. *Coupled longitudinal and bending vibrations of a rotating shaft with an open crack*. Journal of Sound and Vibration, 117(1):81–93, 1987.

[102] C. A. Papadopoulos und A. D. Dimarogonas. *Stability of cracked rotors in the coupled vibration mode*. Journal of Vibration, Acoustics, Stress, and Reliability in Design - Transactions of the ASME, 110(1):356–359, 1988.

[103] C. A. Papadopoulos und A. D. Dimarogonas. *Coupled vibrations of a cracked shaft.* *Journal of Vibration - Transactions of the ASME*, 114(4):461–467, 1992.

[104] T. H. Patel und A. K. Darpe. *Influence of crack breathing model on nonlinear dynamics of a cracked rotor.* *Journal of Sound and Vibration*, 311:953–972, 2008.

[105] P. Pennacchi, N. Bachschmid und A. Vania. *A model-based identification method of transverse cracks in rotating shafts suitable for industrial machines.* *Mechanical Systems and Signal Processing*, 20:2112–2147, 2006.

[106] J. E. T. Penny und M. I. Friswell. *Simplified modelling of rotor cracks.* *Key Engineering Materials*, 245-246:223–232, 2003.

[107] J. E. T. Penny und M. I. Friswell. *The dynamics of rotating machines with cracks.* *Materials Science Forum*, 440-441:311–320, 2003.

[108] R. H. Plaut, R. H. Andruet und S. Suherman. *Behavior of a cracked rotating shaft during passage through a critical speed.* *Journal of Sound and Vibration*, 173(5): 577–589, 1994.

[109] C. Proppe. *Simulation dynamischer Systeme.* Institut für Technische Mechanik, Karlsruher Institut für Technologie, 2010.

[110] G.-L. Qian, S.-N. Gu und J.-S. Jiang. *The dynamic behaviour and crack detection of a beam wuth a crack.* *Journal of Sound and Vibration*, 138(2):233–243, 1990.

[111] W. Ramberg und W. Osgood. *Description of stress-strain curves by three parameters.* *National Advisory Committe For Aeronautics, Washington DC.*, Technical Note No.902, 1943.

[112] J. S. Rao. *History of rotating machinery dynamics.* Springer-Verlag, 2011.

[113] J. S. Rao und R. Sreenivas. *Dynamics of asymmetric rotors using solid models.* In *Proceedings of the International Gas Turbine Congress, Tokyo, Japan*, 2003.

[114] P. F. Rizos, N. Aspragathos und A. D. Dimarogonas. *Identification of crack location and magnitude in a cantilever beam from the vibration modes.* *Journal of Sound and Vibration*, 138(3):381–388, 1990.

[115] H. E. Rosinger und I. G. Ritchie. *On Timoshenko's correction for shear in vibrating isotropic beams.* *Journal of Physics D: Applied Physics*, 10:1461–1466, 1977.

[116] G. Sabnavis, R. G. Kirk und M. Kasarda. *Cracked shaft detection and diagnostics: a literature review.* *Shock and Vibration Digest*, 36(4), 2004.

[117] J. T. Sawicki, X. Wu, G. Y. Baakline und M. I. Friswell. *Dynamic behavior of cracked flexible rotor subjected to constant driving torque.* In *Proceeding of the ISCORMA-2, Gdansk, Poland*, 2003.

[118] I. Scheider. *Derivation of separation laws for cohesive models in the course of ductile fracture. Engineering Fracture Mechanics*, 76:1450–1459, 2009.

[119] I. Scheider und W. Brocks. *Simulation of cup-cone fracture using the cohesive model. Engineering Fracture Mechanics*, 70:1943–1961, 2003.

[120] I. Scheider und W. Brocks. *Cohesive elements for thin-walled structures. Computational Materials Science*, 37:101–109, 2006.

[121] I. Scheider, M. Schödel, W. Brocks und W. Schönfeld. *Crack propagation analyses with CTOA and cohesive model: Comparison and experimental validation. Engineering Fracture Mechanics*, 73:252–263, 2006.

[122] G. Schweitzer. *Magnetic bearings: Theory, design, and application to rotating machinery*. Springer-Verlag, 2009.

[123] A. S. Sekhar. *Vibration characteristics of a cracked rotor with two open cracks. Journal of Sound and Vibration*, 223(4):497–512, 1999.

[124] A. S. Sekhar. *Crack identification in a rotor system: a model-based approach. Journal of Sound and Vibration*, 270:887–902, 2004.

[125] A. S. Sekhar und J. K. Dey. *Effects of cracks on rotor system instability. Mechanism and Machine Theory*, 35:1657–1674, 2000.

[126] A. S. Sekhar und B. S. Prabhu. *Crack detection and vibration characteristics of cracked shafts. Journal of Sound and Vibration*, 157(2):375–381, 1992.

[127] A. S. Sekhar und B. S. Prabhu. *Vibration and stress fluctuation in cracked shafts. Journal of Sound and Vibration*, 169(5):655–667, 1994.

[128] A. S. Sekhar und B. S. Prabhu. *Condition monitoring of cracked rotors through transient response. Mechanism and Machine Theory*, 33:1167–1175, 1998.

[129] A. S. Sekhar und P. B. Prasad. *Dynamic analysis of a rotor system considering a slant crack in the shaft. Journal of Sound and Vibration*, 208(3):457–474, 1997.

[130] C. Shet und N. Chandra. *Analysis of energy balance when using cohesive zone models to simulate fracture processes. Journal of Engineering Materials and Technology*, 124:440–450, 2002.

[131] Y. S. Shih und J. J. Chen. *Analysis of fatigue crack growth on a cracked shaft. International Journal of Fatigue*, 19(6):477–485, 1997.

[132] Y. S. Shih und J. J. Chen. *The stress intensity factor study of an elliptical cracked shaft. Nuclear Engineering and Design*, 214:137–145, 2002.

[133] T. Siegmund und W. Brocks. *Tensile decohesion by local failure criteria. Technische Mechanik*, Band 18, Heft 4:261–270, 1998.

[134] T. Siegmund und W. Brocks. *A numerical study on the correlation between the work of separation and the dissipation rate in ductile fracture.* Engineering Fracture Mechanics, 67:139–154, 2000.

[135] J.-J. Sinou. *Effects of a crack on the stability of a non-linear rotor system.* International Journal of Non-Linear Mechanics, 42:959–972, 2007.

[136] J.-J. Sinou und A. W. Lees. *The influence of cracks in rotating shafts.* Journal of Sound and Vibration, 285:1015–1037, 2005.

[137] J.-J. Sinou und A. W. Lees. *A non-linear study of a cracked rotor.* European Journal of Mechanics-A/Solids, 26:152–170, 2007.

[138] N. G. Stephen und M. Levinson. *A second order beam theory.* Journal of Sound and Vibration, 67(3):293–305, 1979.

[139] K. Takahashi. *An approach to investigate the instability of the multiple-degree-of-freedom parametric dynamic systems.* Journal of Sound and Vibration, 78(4):519–529, 1981.

[140] D. L. Thomas, J. M. Wilson und R. R. Wilson. *Timoshenko beam finite elements.* Journal of Sound and Vibration, 31(3):315–330, 1973.

[141] T. C. Tsai und Y. Z. Wang. *Vibration analysis and diagnosis of a cracked shaft.* Journal of Sound and Vibration, 192(3):607–620, 1996.

[142] A. Turon, P. P. Camanho, J. Costa und C. G. Dávila. *A damage model for the simulation of delamination in advanced composites under variable-mode loading.* Mechanics of Materials, 38:1072–1089, 2006.

[143] A. Turon, J. Costa, P. P. Camanho und C. G. Dávila. *Simulation of delamination in composites under high-cycle fatigue.* Composites Part A, 38:2270–2282, 2007.

[144] A. Turon, C. G. Dávila, P. P. Camanho und J. Costa. *An engineering solution for mesh size effects in the simulation of delamination using cohesive zone models.* Engineering Fracture Mechanics, 74:1665–1682, 2007.

[145] V. Tvergaard und J. W. Hutchinson. *The relation between crack growth resistance and fracture process parameters in elastic-plastic solids.* Journal of the Mechanics and Physics of Solids, 40(6):1377–1397, 1992.

[146] K. Y. Volokh. *Comparison between cohesive zone models.* Communications in Numerical Methods in Engineering, 20:845–856, 2004.

[147] J. Wauer. *On the dynamics of cracked rotors: A literature survey.* Applied Mechanics Review, 43(1):13–17, 1990.

[148] J. Wauer. *Modelling and formulation of equations of motion for cracked rotating shafts.* International Journal of Solids and Structures, 26(8):901–914, 1990.

[149] J. Wauer. *Kontinuumsschwingungen*. Vieweg+Teubner, 2008.

[150] H. L. Wettergren und K.-O. Olsson. *Dynamic instability of a rotating asymmetric shaft with internal viscous damping supported in anisotropic bearings*. Journal of Sound and Vibration, 195(1):75–84, 1996.

[151] J. Wittenburg. *Schwingungslehre: Lineare Schwingungen, Theorie und Anwendungen*. Springer-Verlag, 1996.

[152] C. Wörnle. *Numerische Methoden der Dynamik*. Technische Dynamik, Universität Rostock, 2007.

[153] J. Wu und F. Ellyin. *A study of fatigue crack closure by elastic-plastic finite element analysis for constant-amplitude loading*. International Journal of Fracture, 82:43–65, 1996.

[154] X. Wu und J. Meagher. *A two-disk extended Jeffcott rotor model distinguishing a shaft crack from other rotating asymmetries*. International Journal of Rotating Machinery, Article ID 846365, 2008.

[155] X.-P. Xu und A. Needleman. *Void nucleation by inclusion debonding in a crystal matrix. Modelling and Simulation in Materials Science and Engineering*, 1:111–132, 1993.

[156] T. Yamamoto und Y. Ishida. *Linear and nonlinear rotordynamics: A modern treatment with applications*. John Wiley and Sons, Inc., 2001.

[157] B. Yang und C. S. Suh. *On fault induced nonlinear rotary response and instability*. International Journal of Mechanical Science, 48:1103–1125, 2006.

[158] J. Yang, T. Zheng, W. Zhang, H. Yuan und B. Wen. *The complicated response of a simple rotor with a fatigue crack*. In Proceeding of the 6th IFToMM International Conference on Rotating Machinery, Sydney, Australia, 2002.

[159] J. Zou und J. Chen. *A comparative study on time-frequency feature of cracked rotor by Wigner-Ville distribution and wavelet transform*. Journal of Sound and Vibration, 276:1–11, 2004.

[160] Z. Zou, S. R. Reid und S. Li. *A continuum damage model for delaminations in laminated composites*. Journal of the Mechanics and Physics of Solids, 51:333–356, 2003.